高等职业教育精品示范教材（电子信息课程群）

Android 软件应用与实践

主　编　梁　平　高　峰

副主编　张　扬　李云平　周国亮

参　编　王　超　葛　佳

·北京·

内 容 提 要

本书的内容包括Android简介、Android开发快速入门、Android用户界面开发、Android消息与广播、Android数据存储、Android图形图像及综合案例开发——简易通讯录，以实例、实训项目教学为主，注重实践教学。

理论内容通俗易懂；内容循序渐进、由浅入深，达到逐步提高的目的；用实例解释概念，达到理论与实践紧密结合的目的；每一章均编入了“应用举例”，使读者将本章的概念、实例通过应用举例的方式加深理解，达到实践的目的；每一章均编入了“实训项目”内容，配合授课教师组织实践教学，达到教学的目的。

本教材的编写主要面向高职高专教育，力求实用好学，适合相关专业的学生学习，以及作为其他专业的自学和参考用书。

本书提供电子教案，读者可以从中国水利水电出版社网站和万水书苑上下载，网址为：http://www.waterpub.com.cn/softdown/和 http://www.wsbookshow.com。

图书在版编目（CIP）数据

Android软件应用与实践 / 梁平，高峰主编. -- 北京：中国水利水电出版社，2016.7（2018.12 重印）
高等职业教育精品示范教材. 电子信息课程群
ISBN 978-7-5170-4579-3

Ⅰ. ①A… Ⅱ. ①梁… ②高… Ⅲ. ①移动终端－应用程序－程序设计－高等职业教育－教材 Ⅳ. ①TN929.53

中国版本图书馆CIP数据核字(2016)第173836号

策划编辑：祝智敏　责任编辑：李　炎　加工编辑：韩莹琳　封面设计：李　佳

书　名	高等职业教育精品示范教材（电子信息课程群） Android 软件应用与实践 Android RUANJIAN YINGYONG YU SHIJIAN
作　者	主　编　梁　平　高　峰 副主编　张　扬　李云平　周国亮
出版发行	中国水利水电出版社 （北京市海淀区玉渊潭南路 1 号 D 座　100038） 网址：www.waterpub.com.cn E-mail：mchannel@263.net（万水） sales@waterpub.com.cn 电话：（010）68367658（营销中心）、82562819（万水）
经　售	全国各地新华书店和相关出版物销售网点
排　版	北京万水电子信息有限公司
印　刷	三河市祥宏印务有限公司
规　格	184mm×240mm　16 开本　14.75 印张　326 千字
版　次	2016 年 7 月第 1 版　2018 年 12 月第 2 次印刷
印　数	3001—6000 册
定　价	35.00 元

高等职业教育精品示范教材（电子信息课程群）

丛书编委会

I

序

为贯彻落实国务院印发的《关于加快发展现代职业教育的决定》，加快发展现代职业教育，形成适应发展需求、产教深度融合、中职高职衔接、职业教育与普通教育相互沟通的现代职业教育体系，在围绕中国职业技术教育学会研究课题的基础上联合大批的一线教师和技术人员，共同组织出版"高等职业教育精品示范教材（电子信息课程群）"职业教育系列教材。

职业教育在国家人才培养体系中有着重要位置，以"服务发展"为宗旨，以"促进就业"为导向，适应技术进步和生产方式变革以及社会公共服务的需要，从而培养数以亿计的高素质劳动者和技术技能人才。紧紧围绕国家发展职业教育的指导思想和基本原则，编委会在调研、分析、实践等环节的基础上，结合社会经济发展的需求，设计并打造电子信息课程群的系列教材。本系列教材配合各职业院校专业群建设的开展，涵盖软件技术、移动互联、网络系统管理、软件与信息管理等专业方向，有利于建设开放共享的实践环境，有利于培养"双师型"教师团队，有利于学校创建共享型教学资源库。

本次精品示范系列教材的编写工作遵循以下几个基本原则：

（1）体现以就业为导向、产学结合的发展道路。学科和专业同步加强，按企业需要和岗位需求来对接培养内容。既反映学科的发展趋势，又能结合专业教育的改革，及时反映教学内容和教学体系的调整更新。

（2）采用项目驱动、案例引导的编写模式。打破传统的以学科体系设置课程体系和以知识点为核心的框架，更多地考虑学生所学知识与行业需求及相关岗位、岗位群的需求相一致，坚持"工作流程化""任务驱动式"，突出"走向职业化"的特点，努力培养学生的职业素养和职业能力，实现教学内容与实际工作的高仿真对接，真正以培养技术技能型人才为核心。

（3）专家、教师共建团队，优化编写队伍。由来自于职业教育领域的专家、行业企业专家、院校教师、企业技术人员协同组合编写队伍，跨区域、跨学校进行交叉研究、协调推进，把握行业发展和创新教材发展的方向，融入专业教学的课程设置与教材内容。

（4）开发课程教学资源，推进专业信息化建设。从充分关注人才培养目标、专业结构布局等入手，开发补充性、更新性和延伸性教辅资料，开发网络课程、虚拟仿真实训平台、工作过程模拟软件、通用主题素材库以及名师讲义等多种形式的数字化教学资源，建立动态、共享的课程教材信息化资源库，服务于系统培养技术技能型人才。

电子信息类教材建设是提高电子信息领域技术技能型人才培养质量的关键环节，是深化职业教育教学改革的有效途径。为了促进现代职业教育体系建设，使教材建设全面对接教学改革和行业需求，更好地服务区域经济和社会发展，我们殷切希望各位职教专家和老师提出建议，并加入到我们的编写队伍中来，共同打造电子信息领域的系列精品教材！

丛书编委会

2014 年 6 月

前　言

本书从初学者的角度出发，通过通俗易懂的语言及丰富多彩的项目实例，详细介绍了 Android 应用程序开发应该掌握的各方面技术。通过阅读本书，读者不仅能够了解与 Android 系统相关的理论知识，还可以掌握当下最热门的 Android 应用程序的编写技巧。

本书针对高职和成人院校的教学特点，本着理论够用、以实践技能培养为主的原则，注重对读者动手能力的培养，编入了大量的实例及实训项目的内容，帮助读者更好地进行 Android 应用程序开发设计与实践。

全书共分 7 章，按照循序渐进的原则，详细介绍了 Android 应用程序开发技术，给出内容全面、步骤完整的操作实例。第一章介绍了 Android 的起源及系统架构；第二章介绍了 Android 开发环境的搭建、AVD 的创建、Android 项目的创建及运行、Activity 的简单介绍；第三章详细介绍了 Android 用户界面开发，包括界面布局设计、常用界面组件、界面资源的定义与使用、单选按钮及相关事件、本单元的实训项目；第四章介绍了 Android 消息与广播，包括 Intent 与 Activity、BroadcastReceive 广播组件应用及本单元的实训项目；第五章介绍了 Android 数据存储，包括 SharedPreferences 存储、File 文件存储、SQLite 数据库、ContentProvider 使用及本单元相关实训项目；第六章介绍了 Android 图形图像及动画、相关实训项目等；第七章为综合案例开发设计详解。

本书由多位从事高等院校移动互联专业教学的一线老师，根据多年的教学和项目开发总结编写而成，本书的主要特点可归纳如下：

1．理论内容通俗易懂。力求用通俗易懂的语言解释较为复杂的实例操作步骤，即使非计算机专业的读者也能理解，达到会应用的目的。

2．内容循序渐进、由浅入深，达到逐步提高的目的。

3．用实例解释概念，达到理论与实践紧密结合的目的。

4．每一单元均编入了大量的实例及实训项目，使读者将单元内的概念、实例通过应用举例的方式加深理解，达到实践的目的。

本书可作为高等院校“Android应用开发”课程教学用书，也可作为专升本培训教材。

书中难免有疏漏之处，敬请各位读者提出宝贵意见，以便我们及时修正。

编　者

2016年5月

III

目　录

1 Android 简介

Android 是一种基于 Linux 的自由及开放源代码的操作系统，广泛用于移动设备上，如智能手机和平板电脑，由 Google 公司和开放手机联盟领导及开发。Android 手机操作系统目前已经成为成熟的手机操作系统之一。随着 Android 的高速发展，将来会有越来越多的开发者加入进来。为了和更多开发者共同探索 Android 的多彩世界，我们将通过本章的学习，了解 Android 平台的发展历史，掌握 Android 的特征和 Android 平台的体系结构。

1.1 手机操作系统

手机操作系统主要应用在智能手机上。主流的智能手机有 Google 的 Android 和苹果的 iOS 等。智能手机与非智能手机都支持 JAVA，智能机与非智能机的区别主要看能否基于系统平台的功能扩展，非 JAVA 应用平台，还有就是支持多任务。

常见手机系统的比较

1. Symbian

Symbian 是一个实时的、多任务的纯 32 位操作系统，具有功耗低、内存占用少等特点，在有限的内存和运存情况下，非常适合手机等移动设备使用，经过不断发展完善，可以支持 GPRS、蓝牙、SyncML、以及 3G 技术。它包含联合的数据库、使用者界面架构和公共工具的参考实现。由于缺乏新技术支持，Symbian 的市场份额日益萎缩。2013 年 1 月 24 日，诺基亚宣布，今后将不再发布 Symbian 系统的手机，意味着 Symbian 这个智能手机操作系统，在长达 14 年的历史之后，终于迎来了谢幕。

2. BlackBerry

BlackBerry（黑莓）无疑给用户提供了高容量、易操作的移动平台。黑莓系统提供了强大的邮件处理与加密功能，并且其加密方式不公开，保证了个人账户的安全性，也因此，黑莓成为了企业业务用户的不二选择，但是其娱乐性、普及性与易用性远不如 Android 与 IOS。致使黑莓目前只是适用于小范围用户的移动设备操作系统。

3. IOS

苹果 IOS 系统是目前引领潮流的操作系统之一，IOS 系统注重交互体验、触控、滑动、轻触开关等的体验，苹果更是将其发挥到了极致，但是，IOS 对于开发者来说又是一个让人又爱又恨的系统。苹果相对于 Android 来讲较为封闭，Android 为完全开源的系统，在开发过程中留给开发者的发挥空间非常大，但是苹果在开发过程中存在着种种限制，导致在 IOS 应用系统的开发过程中，开发者在样式、功能实现方式上都不能自由发挥。

4. Windows Phone

2008 年，在 Android 和 IOS 的冲击下，Windows 重组了 Windows mobile 的小组。Windows Phone 的风格采用扁平化设计 UI，初始界面采用动态转的元素，整体看起来简洁清新，整体有 Win8 系统的特点，动态转可以连接至应用程序。其内置的 Office 办公套件和 Outlook 更是使用户的办公更加方便快捷。在开发方面，Windows phone 同样提供了很好的开发工具，但是微软也制定了很多规范，并且要求开发者严格遵循这些开发规范。Windows Phone 相对而言还是一个年轻的操作系统、在娱乐性，应用数量方面和 Android 还有一定差距。

5. Android

最后，我们来谈一谈 Android。Android 从问世之后就开始了非常快速的发展，直到今天 Android 已成为市场上主流的移动设备操作系统。Android 发展之快要得益于它开源的做法，它使得全世界所有的 Android 爱好者都可以自由的对系统进行优化，或者为其编写应用，它的开源程度远高于其他手机操作系统，这让移动开发者在开发过程中享有更多的自由空间，可以通过最有效、最快捷的方式来实现期望的功能，刺激了全世界编程爱好者的开发热情。同样，这也减轻了厂家的压力与成本，使 Android 系统能够进一步成长。

1.2 Android 起源

1.2.1 开放手机联盟

开放手机联盟（Open Handset Alliance，OHA）由谷歌公司于 2007 年发起的一个全球性的联盟组织，成立时包含 34 家联盟成员，现在已经增加到 50 家。

（1）联盟组织的目标：研发移动设备的新技术，用以大幅削减移动设备开发与推广成本；同时通过联盟各个合作方的努力，建立移动通信领域新的合作环境，促进创新移动设备的开发，创造了目前移动平台实现的用户体验。

开放手机联盟成员，如图 1-1 所示。

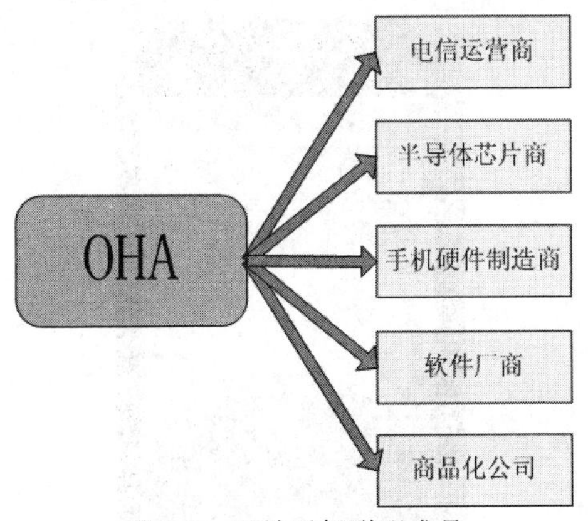

图 1-1　开放手机联盟成员

（2）电信运营商：中国移动通信、KDDI（日本）、NTT DoCoMo（日本）、Sprint Nextel（美国）、T-Mobile（美国）、Telecom（意大利）、中国联通、Softbank（日本）、Telefonica（西班牙）和 Vodafone（英国）。

（3）半导体芯片商：Audience（美国）、AKM（日本）、ARM（英国）、Atheros Communications（美国）、Broadcom（美国）、Intel（美国）、Marvell（美国）、nVIDIA（美国）、Qualcomm（美国）、SiRF（美国）、Synaptics（美国）、ST-Ericsson（意大利、法国和瑞典）和 Texas Instruments（美国）。

（4）手机硬件制造商：Acer（中国台湾）、华硕（中国台湾）、Garmin（中国台湾）、宏达电（中国台湾）、LG（韩国）、三星（韩国）、华为（中国）、摩托罗拉（美国）、索尼爱立信（日本和瑞典）和东芝（日本）。

（5）软件厂商：Ascender Corp（美国）、eBay（美国）、谷歌（美国）、LivingImage（日本）、NuanceCommunications（美国）、Myraid（瑞士）、Omron（日本）、PacketVideo（美国）、SkyPop（美国）、Svox（瑞士）和 SONiVOX（美国）。

（6）商品化公司：Aplix Corporation（日本）、Noser Engineering（瑞士）、Borqs（中国）、TAT-The Astonishing（瑞典）、Teleca AB（瑞典）和 Wind River（美国）。

1.2.2　Android 起源

起初，Andy Rubin等人创建了 Android 公司，并组建 Android 团队。之后Google低调收购了成立仅 22 个月的高科技企业 Android 及其团队。Andy Rubin 继续负责 Android 项目。

2007 年 11 月 5 日，Android 操作系统正式问世，同日谷歌宣布建立一个全球性的联盟组织，并与 84 家硬件制造商、软件开发商及电信营运商组成开放手持设备联盟来共同研发改良 Android 系统。在 2008 年 9 月，谷歌正式发布了 Android 1.0 系统，这也是 Android 系统最早的版本，如图 1-2 所示。

图 1-2 Android 1.0 系统

Google 在完善了虚拟键盘之后推出了 Android 1.5，并且从这个版本之后，Google 开始使用甜品的名字来命名 Android，如 1.5 Cupcake，1.6 Donut。

2010 年，Android 的驱动程序从 Linux 内核状态树上除去，从此，Android 离开了 Linux 自成一家。并在同年 5 月份，Android 2.2 Froyo 发布。

到 2011 年 7 月，Android 系统设备的用户总数达到了 1.35 亿，Android 系统已经成为智能手机领域占有量最高的系统。

1.2.3 Android 的未来

Android 采用 WebKit 浏览器引擎，具备触摸屏、高级图形显示和上网功能，用户可以在手机上收发邮件、浏览网页等。

就硬件来讲，目前主流架构就是 ARM 与英特尔，Android+ARM 已经足够优秀，英特尔在看到移动市场的巨大利润之后，后悔当初放弃涉足移动市场的决定，但是现在英特尔已经不能在保证利润的前提下，压低价格同 Android 与 ARM 的组合来竞争市场，而苹果则会继续在垂直一体化的非移动设备上赢取利润。

在本地商务方面，Android 更是在努力向前推进，小至公交车的一次刷卡，大到商场购物，只需扫描手机，便可完成支付把商品拿走。

同时，在物联网高速发展的今天，Google 也正在将 Android 系统用于智能家居控制，例如 Android 手机可以当做 Google TV 的遥控器，Android 很有可能在不久的将来成为物联网时代的主导系统。

1.3 Android 系统的框架

1.3.1 架构总览

从 Google 开源资料可以得知，Android 系统架构由五部分组成，分别是：Linux Kernel、Android Runtime、Libraries、Application Framework 和 Applications，如图 1-3 所示。

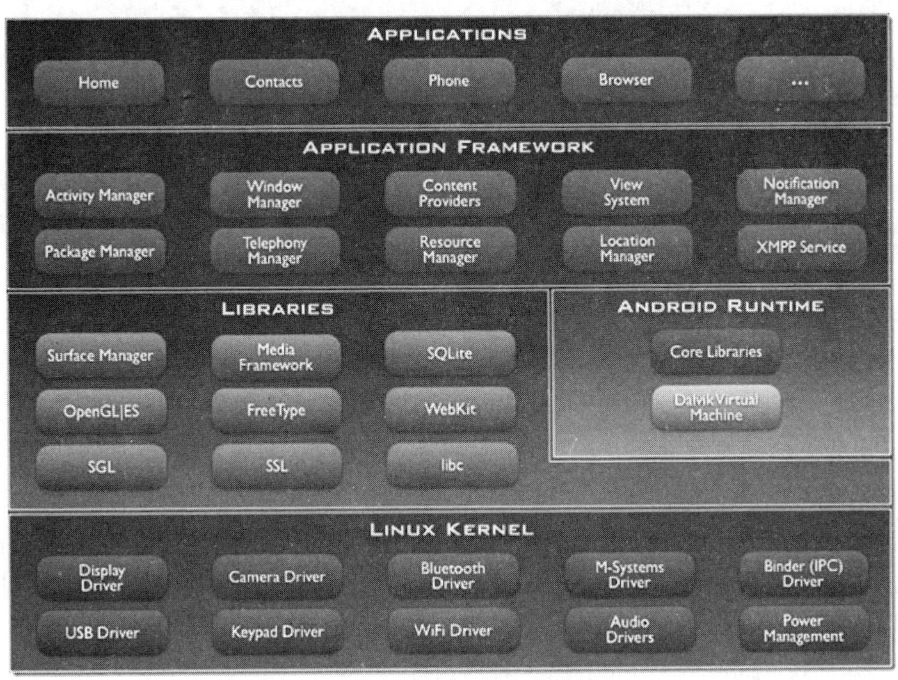

图 1-3　Android 系统架构

1.3.2 Android 架构详解

Linux Kernel

Android 是基于 Linux 2.6 内核提供核心系统服务，但并不是生搬硬套，而是对内核做了部分的修改，在此基础之上，Android 系统核心可提供安全、内存管理、进程管理、网络堆栈、驱动模型等服务。同时 Linux Kernel 也作为硬件和软件之间的抽象层，能够隐藏具体硬件细节而为上层提供统一的服务。

实现的主要功能有：

- 进程的调度管理：进程的创建与销毁、进程间通信、协调进程间资源分配、避免死锁等问题。

- 内存管理：对可用内存统一管理，实现内存的定位、分配与回收。
- 驱动模型：完成与硬件的通信。

Android Runtime

Android Runtime 由两部分组成，分别是核心库与虚拟机。

核心库可提供 Java 语言核心库中大多数功能。Dalvik 虚拟机是基于寄存器实现的，它可以执行的文件格式是.dex。SDK 中包含了转换工具 dx，可以把.class 文件转化成.dex 格式，供虚拟机执行。

Libraries

Android 中包含了一部分 C/C++库，可用于 Android 中的不同组件，他们可以通过程序框架为开发者提供服务，以下为一部分核心库：

- 系统 C 库（libc）：一个从 BSD 继承的标准 C 系统函数库。
- SGL：2D 图形引擎，用于 2D 图形渲染。
- Open GL ES：可以使用 3D 硬件加速，用于 3D 图形渲染。
- Free Type：位图和矢量字体的显示。
- Surface Manager：管理显示子系统并为多个程序提供 2D、3D 的无缝融合。
- Media Framework：基于 Packet Video Open CORE，支持多种视频格式，以及常用 DOM 操作，同时支持静态图像文件。
- Lib Web Core：浏览器引擎，用于驱动 Android 浏览器以及程序内嵌的 Web 界面。

Application Framework

Application Framework 指一个程序正常运行所需的全部组件，这使得开发者能够自由地开发自己的程序，添加状态栏，制作 Widget，设计主题等。同样，开发者也可以访问核心应用程序所使用的框架 APIs，它简化了组建重用，使任何应用程序都能发布它的功能块或者使用它所发布的功能块。同时，它还支持替换程序组件。

所有应用程序的本质是一系列的服务和系统，包括：

- Activity Manager：管理应用程序的生命周期，并提供常用的导航回退。
- Resource Manager：提供对本地资源的访问，如图形，字符串，布局等。
- Content Providers：是应用程序间实现数据共享，自己能访问别人的数据，或者让别人能访问自己的数据。
- Notification Manager：使应用程序可以在状态栏发布状态或者警告。
- View：一个视图集，包括程序常用的 button，textfield 等。

Applications

同系统同时发布的核心应用程序与常用程序，包括电话、短信、通讯录、照相、电子邮件、浏览器等。全部程序均由 Java 语言编写。

1.4 本章小结

本章内容主要是为了帮助广大读者了解 Android，并为 Android 开发做初步准备。下面我们来回顾一下本章内容：

2007 年，Android 系统正式问世。到目前，Android 已经成为最受欢迎的移动操作系统之一。

Android 的飞速发展得益于 Google 将其开源，并组织联盟对其进行开发完善。

Android 基于 Linux 操作系统，由 Linux Kernel、Android Runtime、Libraries、Application Framework、Applications 等组成。

1.5 本章习题

1．Android 系统相对于其他操作系统的优势是什么？

2．Android 与 Linux 是什么关系？

3．移动操作系统与 PC 系统的区别是什么？

4．Android 开发过程中，Java 语言扮演着什么样的角色？

5．Android 系统包括哪些部分？

2 Android 开发快速入门

学习目标：

对 Android 快速入门，能够开发运行简单 Android 应用，为以后深入学习打下坚实基础。

【知识目标】

- 理解 Android 相关的基本概念
- 熟练搭建 Android 开发运行环境
- 编写一个 Android 应用程序
- 了解 Android 应用四个主要组件

【技能目标】

- 能熟练搭建 Android 开发环境
- 编写运行一个 Android 应用

Android 是一个开放的手机操作系统平台，自 2005 年 8 月由 Google 收购注资至今，已经成为成熟的手机操作系统之一。为移动设备提供了一个包含操作系统、中间件及应用程序的软件叠层架构。2012 年的全球 Android 开发者数量已达 100 万，将来还会有越来越多的开发者加入进来。本章主要讲解如何配置 Android 开发环境，首先介绍 Android 开发所需要的开发包和工具，以及获得它们的方式；其次介绍如何正确安装和配置这些开发包；最后为了测试开发环境，创建一个 Android 项目——Hello Android，并在模拟器上运行和调试该程序。此外还简单介绍开发过程中可能会使用到的各种工具，来帮助大家进一步了解 Android。

2.1 开发环境的搭建

在开始 Android 开发之旅启动之前，首先要搭建环境，Android 开发环境的安装和配置是开发 Android 应用程序的第一步，也是深入 Android 平台的一个非常好的入口。

2.1.1 开发准备工作

配置 Android 开发环境之前，首先需要了解 Android 开发对操作系统的要求，Windows、Mac OS、Linux 操作系统上都可以搭建 Android 开发环境，本书使用 Windows 7 为例进行讲解。先前配置开发环境需下载的开发工具较多，配置也较为繁琐。Google 为使搭建 Android 开发环境变得更简单快捷，把 Eclipse、Android SDK、Android ADT 这三种工具整合为 Android Developer Tools（以下简称 ADT），并把它提供给开发者下载。Android 开发所需的软件版本及其下载地址见表 2-1。

表 2-1　Android 开发工具的版本及下载地址

软件名称	所用版本	下载地址
JDK	1.7	http://www.oracle.com/technetwork/java/javase/downloads/jdk7-downloads-1880260.html
adt-bundle-windows-x86-20130514	20140514	本教材配套光盘中提供

2.1.2 开发工具的安装和使用

1. 安装 JDK

到上一节提供的下载地址将 JDK 进行下载后，开始安装。双击安装文件“jdk-7u75-windows-i586”开始安装。按照提示完成安装即可，安装后生成一个“jdk1.7”的文件夹，如图 2-1 所示。文件夹中的内容如图 2-2 所示。

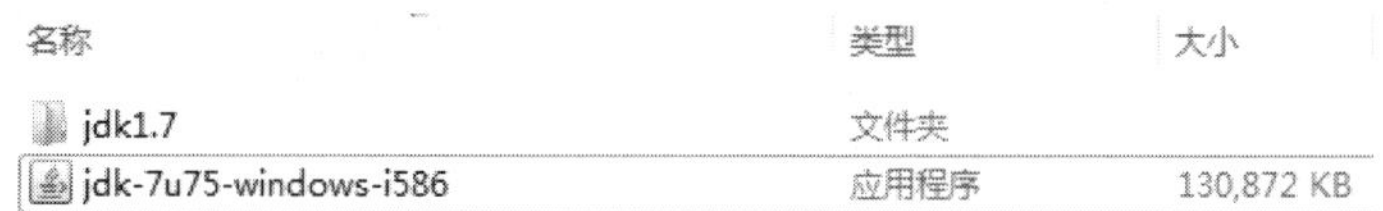

图 2-1　JDK 安装文件及安装后文件夹

2. 配置 JDK 环境变量

（1）右键单击“计算机”，选择“属性”，在弹出“计算机基本信息”对话框的左侧，单击“高级系统设置”，如图 2-3 所示。

（2）弹出“系统属性”对话框，如图 2-4 所示。单击“高级”选项卡中的 环境变量(N)... 按钮，出现“环境变量”设置界面，如图 2-5 所示。

名称	类型	大小
bin	文件夹	
db	文件夹	
include	文件夹	
jre	文件夹	
lib	文件夹	
COPYRIGHT	文件	4 KB
LICENSE	文件	1 KB
README	360 se HTML Do...	1 KB
release	文件	1 KB
src	好压 ZIP 压缩文件	20,296 KB
THIRDPARTYLICENSEREADME	文本文档	173 KB
THIRDPARTYLICENSEREADME-JAVAFX	文本文档	110 KB

图 2-2　JDK 文件夹

图 2-3　“计算机基本信息”对话框

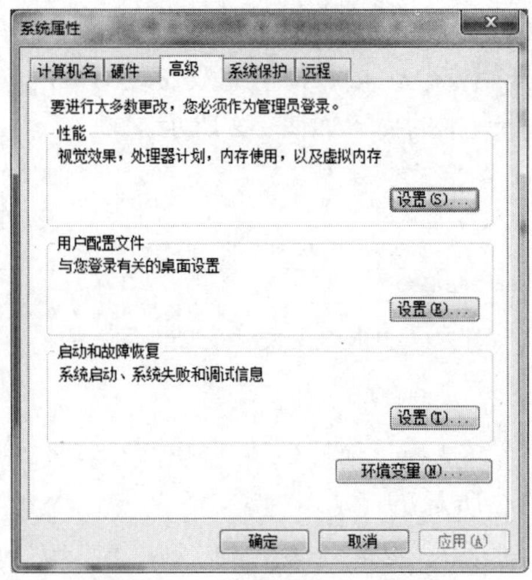

图 2-4　“系统属性”对话框

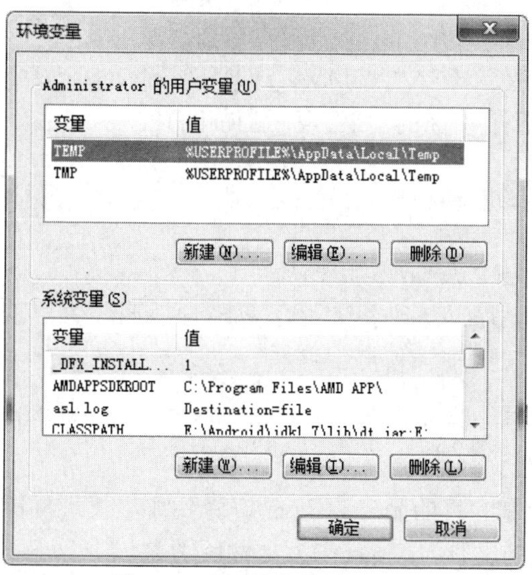

图 2-5　“环境变量”对话框

（3）配置 JAVA_HOME 变量。在下面的“系统变量”处选择新建，在变量名处输入“JAVA_HOME”，变量值处输入之前 JDK 的安装目录，如“E:\Android\jdk1.7”，如图 2-6 所示。

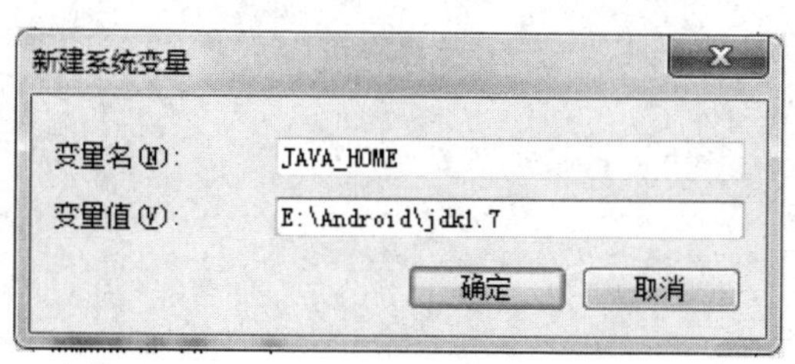

图 2-6　“新建 JAVA_HOME 变量”对话框

配置 JAVA_HOME 变量的目的是得到 JDK 的地址引用，避免每次都要输入很长的地址路径。

（4）配置 Path 路径。在“系统变量”列表中选中 Path 变量，单击编辑按钮，在原有 Path 路径值的末尾添加上一个分号以及安装 JDK 的 bin 文件夹目录，如图 2-7 所示。

编辑系统变量
变量名(N):　Path
变量值(V):　:E\Core-Static;E:\Android\jdk1.7\bin
确定　取消

图 2-7　“编辑 Path 变量”对话框

（5）配置 CLASSPATH 变量。在下面的“系统变量”处选择新建，在“变量名”处输入“JAVA_HOME”，在“变量值”处输入“.;%JAVA_HOME%/lib/rt.jar;%JAVA_HOME%/ lib/tools.jar”，如图 2-8 所示。注意：变量值前面的“.;”表示当前路径。

图 2-8 “新建 CLASSPATH 变量”对话框

配置 CLASSPATH 变量的目的是告诉 Java 编译到哪里发现标准类库。标准类库是已经完成好的可供编程人员利用的文件，以.jar 作为文件后缀名。

（6）最后在 DOS 命令状态下输入“java -version”，查看 JDK 的版本信息，如果显示如图 2-9 中所提示的信息，则说明 JDK 安装成功。

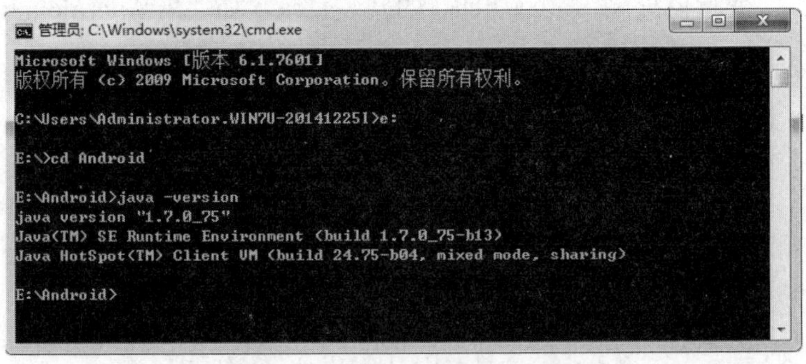

图 2-9 “测试 JDK”对话框

3. ADT 的安装

ADT 工具无需执行安装，在本教材提供的光盘素材中找到“adt-bundle-windows-x86-20130514”压缩文件，对其进行解压后就可以用，不过一定要安装好 JDK。

解压完成后得到“eclipse”“sdk”文件夹和“SDK Manager.exe”可执行文件，如图 2-10 所示。

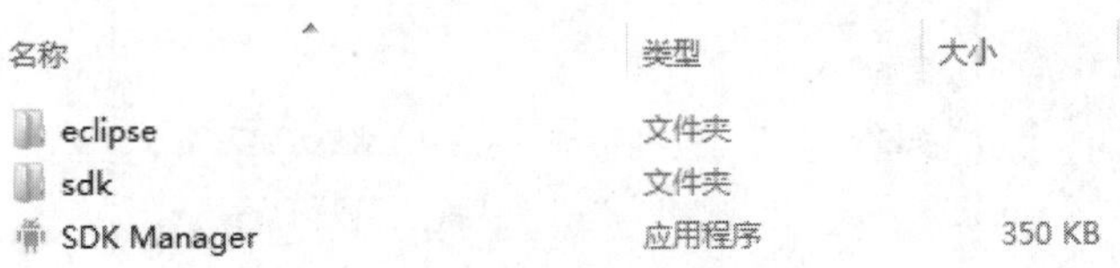

图 2-10 ADT 解压后文件夹

4. 启动 ADT 主界面

（1）打开“eclipse”文件夹，找到“eclipse.exe”可执行文件，双击打开，启动界面如图 2-11 所示。

图 2-11 Eclipse 启动界面

（2）当启动 ADT 时，会弹出如图 2-12 所示的对话框，在此单击 Browse... 可以设置项目的保存路径。下次打开 ADT 如不需再次弹出此对话框，可选中“Use this as the default and do not ask again”前的复选框。

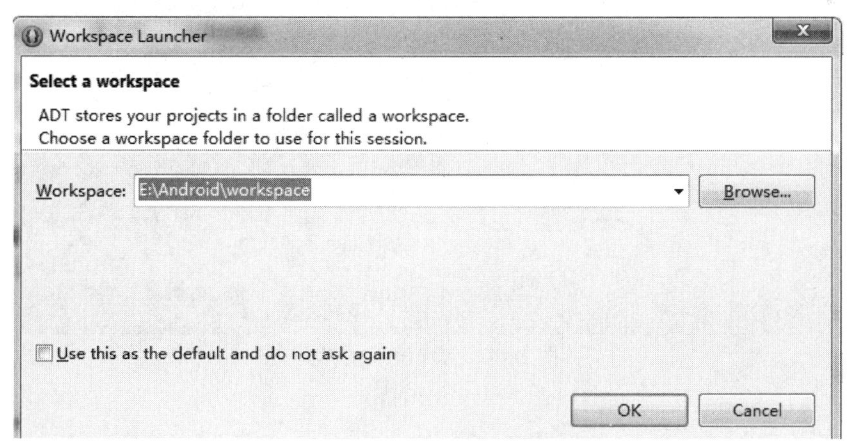

图 2-12 选择 Eclipse 的工作空间

（3）单击 OK 按钮，进入 ADT 的欢迎界面，如图 2-13 所示。

（4）单击“Android IDE”选项卡右边的“×”，关闭欢迎界面，将启动 ADT 的开发环境的主界面，如图 2-14 所示。

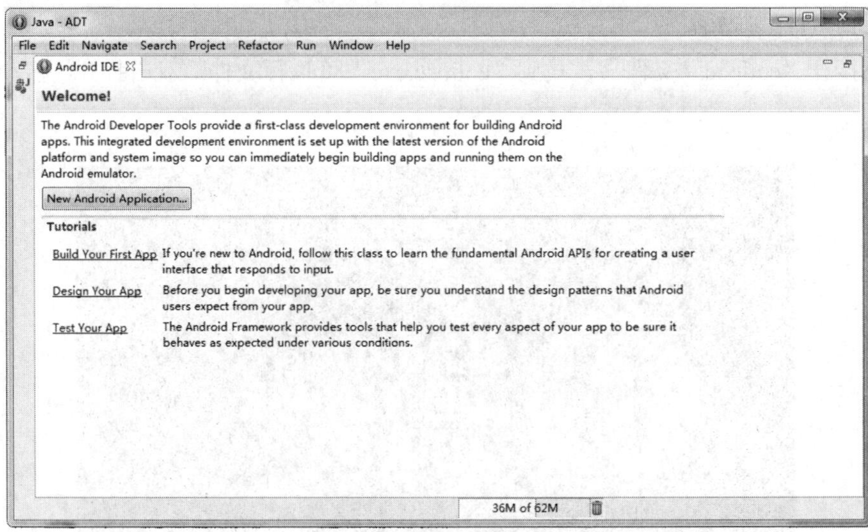

图 2-13　ADT 欢迎界面

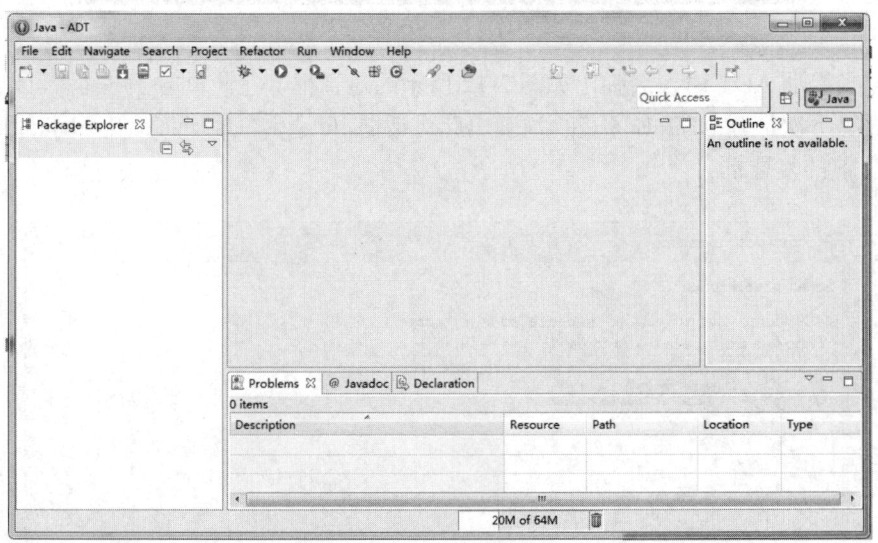

图 2-14　ADT 开发主界面

2.2　创建 AVD

当完成 Android 开发环境搭建后，就可以进行 Android 程序开发。在编写 Android 应用程序时，需要不断测试程序运行结果，为了方便设计人员测试开发程序，Android SDK 提供了 Android 虚拟设备模拟器（Android Virtual Devices），即 AVD，主要目的是方便程序开发人员模拟在真实环境中运行程序一样。为使 Android 应用程序可以在模拟器上运行，在编写程序之前，必须先创建 AVD。

2.2.1 AVD 的操作简介

1. 利用 Eclipse 创建并启动 AVD

（1）启动 Eclipse，单击“Windows”，在下拉菜单中选择“Android Virtual Device Manager”，如图 2-15、图 2-16 所示。

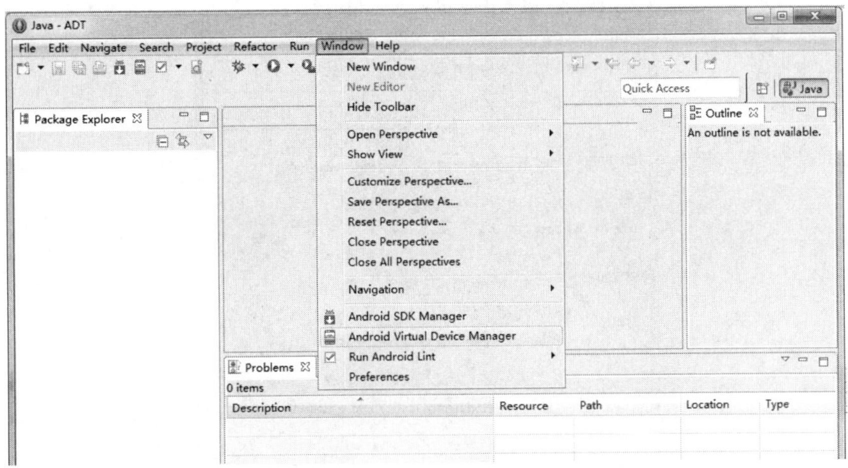

图 2-15　创建 AVD 的 Windows 下拉菜单

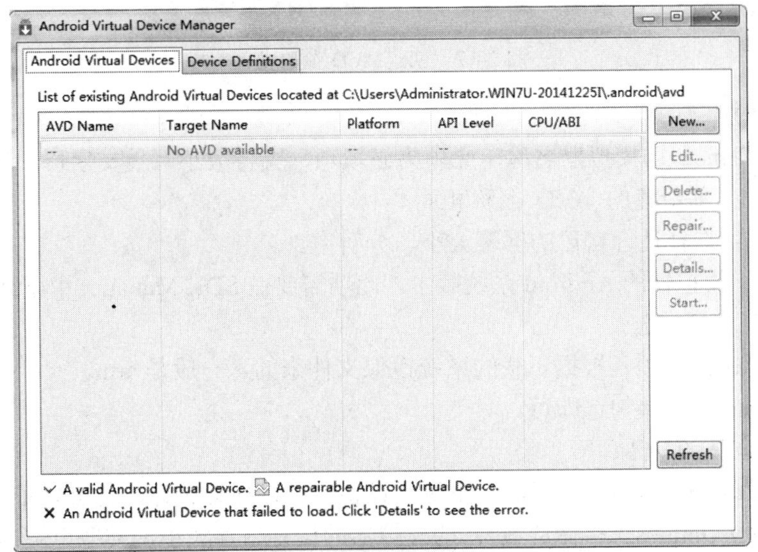

图 2-16　创建 AVD 界面

（2）在默认“Android Virtual Device”选项卡右侧单击 New... 按钮，创建模拟器，如图 2-17 所示。

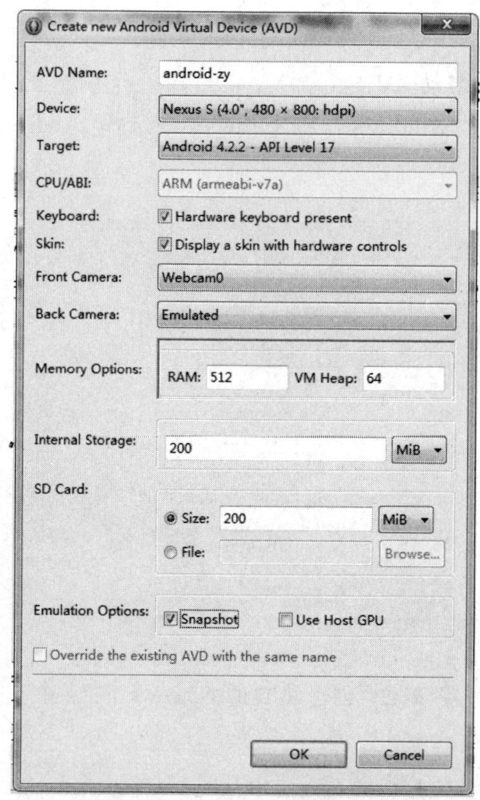

图 2-17　创建 AVD 参数设置

AVD 参数说明：

- AVD Name：即模拟器名称，此项为必填项，支持大小写英文字母、数字、下划线，不能与之前新建的 AVD 名称相同。
- Device：选择适合自己的屏幕大小、分辨率。
- Target：模拟器的 Android 系统版本，此项需要在 SDK Manager 中下载对应系统版本的平台才可以选。
- CPU/ABI：此项是根据下载的系统镜像文件来的，一般是 arm。
- Keyboard：直接勾选即可。
- Skin：直接勾选即可。
- Front/Back Camera：选择前后摄像头设备，可以任意选择，也可以不选。
- Memory Options（存储选项）：即模拟器的运行内存大小，类似电脑内存，可在设置→应用程序中，查看正在运行标签页下显示的具体值。

- Internal Storage（内部存储）：即手机自带存储大小，是模拟器内置存储空间大小，用于存放安装程序和数据的，可在设置→应用程序中，查看其他标签页下显示的具体值。“VM heap”是设置 VM 缓存堆栈的大小，一般使用默认值就可以。
- SD Card：即 SD 存储卡大小，可以选择右侧的下拉选项以改变数值的存储单位，还可以从已有的文件中选择 SD 卡。
- Emulation Options：其他选项可以保持默认，勾选“Snapshot”表示开启快照功能，勾选“Use Host GPU”即表示使用主机的 GPU。

（3）单击 OK 按钮即可完成创建 AVD，在弹出的窗口中就会显示刚刚新建的模拟器，如图 2-18 所示。

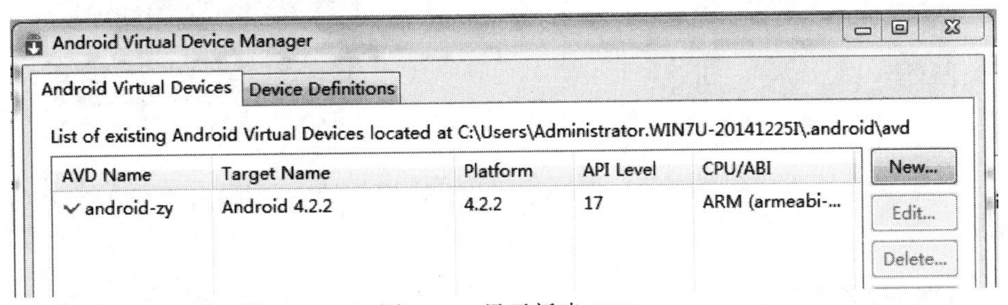

图 2-18　显示新建 AVD

（4）选中该模拟器，单击右侧的 Start... 按钮，可以启动模拟器，如图 2-19 所示，接下来就可以进行 Android 程序开发了。

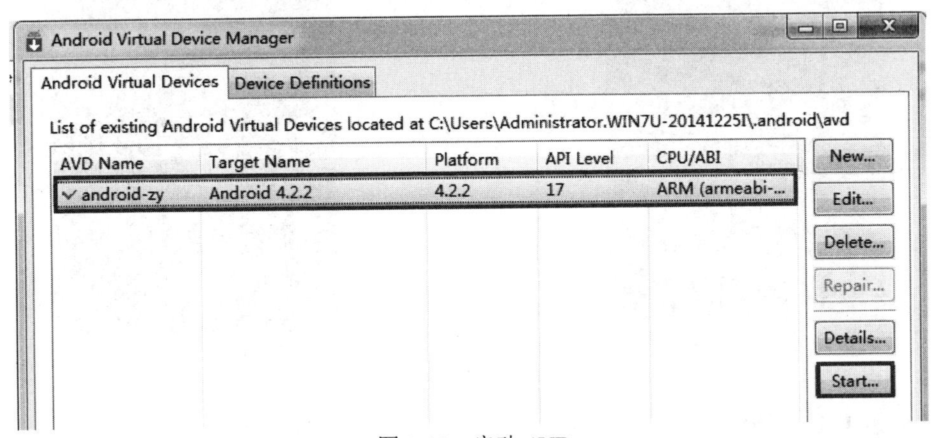

图 2-19　启动 AVD

2. AVD 主界面基本操作

（1）显示主界面。AVD 启动后，会自动显示其主界面，如图 2-20 所示。

（2）查看应用程序。单击屏幕下方的，可显示当前所有的应用程序，如图 2-21 所示。

图 2-20　AVD 主界面

图 2-21　AVD 应用程序界面

（3）查看菜单。单击如图 2-21 所示界面右侧的图标，可显示菜单项 Manage apps 和 System settings，分别选择菜单，可进入对应界面，如图 2-22、图 2-23 所示，进行各种管理与设置。

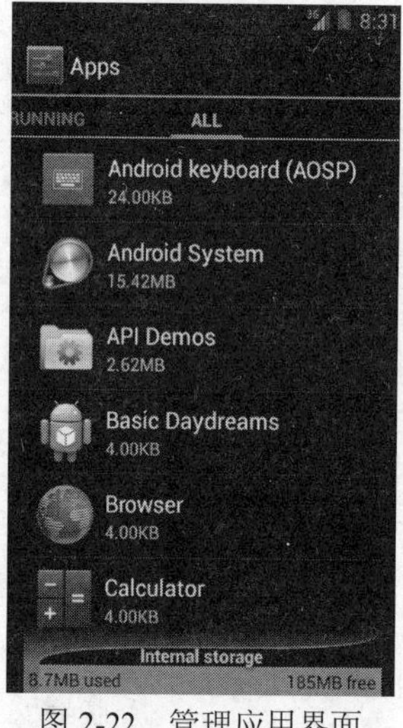

图 2-22　管理应用界面

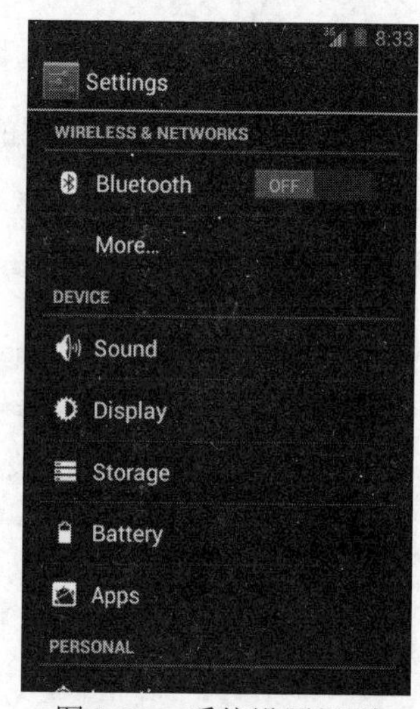

图 2-23　系统设置界面

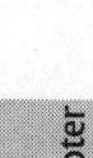

2.2.2　adb shell 命令的使用

1. adb 介绍

SDK 的 platform-tools 文件夹下包含着 Android 模拟器操作的重要命令 adb，adb 的全称为 Android Debug Bridge，也就是调试桥的意思。通过 adb 可以在 Eclipse 中用 DDMS 来调试 Android 程序。借助这个工具，可以管理设备或手机模拟器的状态。还可以进行以下的操作：

- 快速更新设备或手机模拟器中的代码，如应用或 Android 系统升级；
- 在设备上运行 shell 命令；
- 管理设备或手机模拟器上的预定端口；
- 在设备或手机模拟器上复制或粘贴文件。

2. adb 常用命令

（1）命令格式：android list targets。

含义：显示系统中全部 Android 平台。

（2）命令格式：android list avd。

含义：显示系统中全部 AVD。

（3）命令格式：android create avd --name 名称 --target 平台编号。

含义：创建 AVD。

（4）命令格式：emulator -avd 名称 -sdcard ~/名称.img (-skin 1280x800)。

含义：启动 AVD。

（5）命令格式：android delete avd --name 名称。

含义：删除 AVD。

（6）命令格式：adb devices。

含义：显示当前运行的全部模拟器。

（7）命令格式：adb -s 模拟器编号 命令。

含义：对某一模拟器执行命令。

（8）命令格式：adb install -r 应用程序.apk。

含义：安装应用程序。

（9）命令格式：adb help。

含义：查看 adb 命令帮助信息。

2.3 第一个 Android 程序

ADT 提供了简单的生成 Android 应用框架的功能，本节使用 ADT 通过 Eclipse 创建第一个 Android 项目。在 Eclipse IDE 开发环境中建立一个 Android 应用程序之前，首先要创建一个 Android 项目工程，并且建立一个启动配置，建立项目工程的目的就是为开发的应用程序搭建好运行环境需要的支持。然后才可以编写、调试和运行应用程序。

2.3.1 创建 Android 项目

创建 Android 项目的步骤如下：

（1）启动 Eclipse 开发工具，新建一个项目，在弹出的“New Project”对话框的列表中展开“Android”项，然后选择“Android Application Project”子项，如图 2-24 所示。

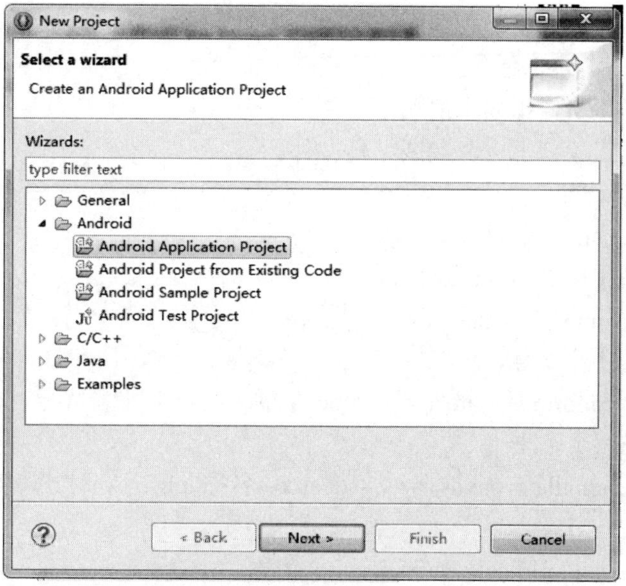

图 2-24　新建一个 Android 工程

（2）单击“Next”按钮，在“Application name”文本框中输入这个应用程序的名字，如 HelloAndroid，在“Project name”文本框中输入工程名称，如 helloandroid，在“Package name”文本框中输入应用程序包的名字，如 com.app.hello，如图 2-25 所示。

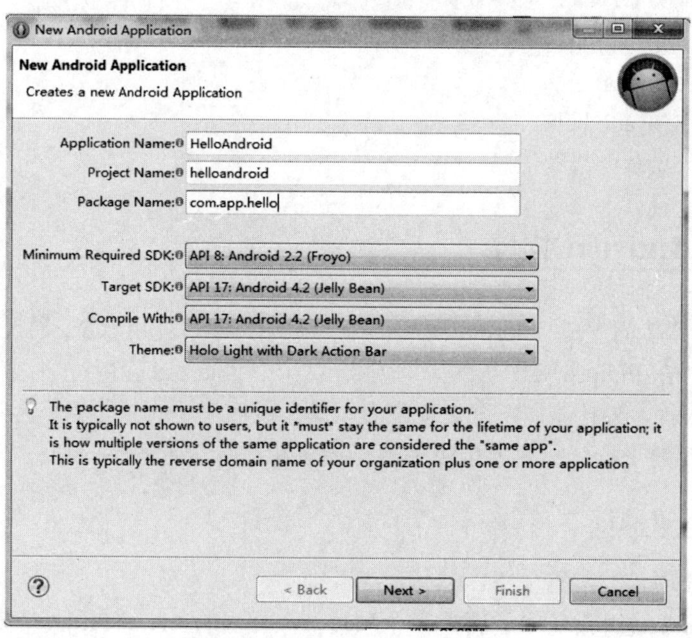

图 2-25　新建 helloandroid 工程

注：Application name 即应用程序呈现给用户的名称。Package name 即 Activity 类的包名称。

（3）连续单击“Next”按钮，对新弹出的对话框均不做改动，选取默认值，最后单击“Finish”按钮，完成 Android 项目的创建，这时 Eclipse 开发平台左边的“Package Explorer”窗口中会显示新创建的项目"helloandroid"，如图 2-26 所示。

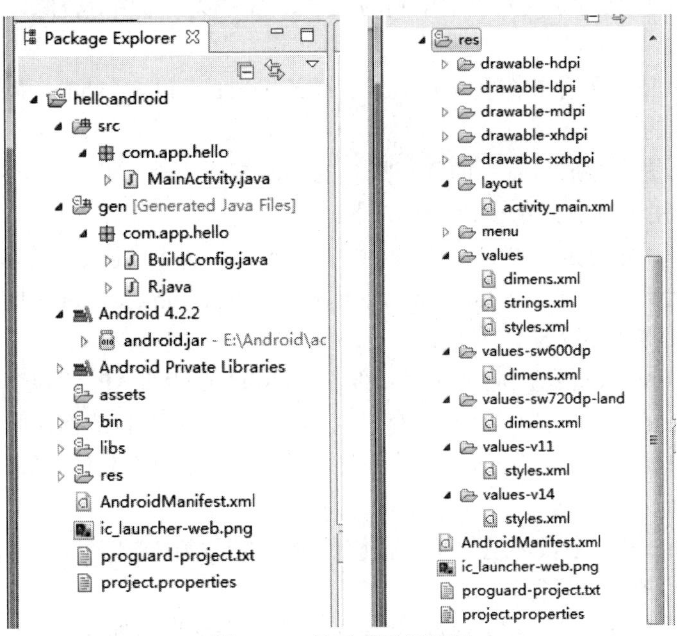

图 2-26　显示项目管理器

此时“helloandroid”项目已经创建好，而且这个项目是由 ADT 插件自动生成，所以不用编写代码，即可运行。

2.3.2　项目框架解析

1．项目结构

在图 2-26 中，可以看到一个 Android 项目的基本目录结构，下面来讲解一下 Android 项目中常见的目录结构。

src/——源代码存放目录，存放项目中的 Java 源程序文件。用于管理由 ADT 自动生成的 Activity 框架代码以及用户自己创建的代码，允许用户修改的 Java 文件都存放在这里。其中 com.app.hello/为程序包目录，这里根据前面新建项目的时候填写的“Package Name”选项来生成应用程序包的 Namespace 命名空间。

gen/——自动生成目录，存放所有由 Android 开发工具自动生成的文件。目录中最重要的就是 R.java 文件，这个文件由 Android 开发工具自动生成，用于引用资源文件夹中的资源文件。ADT 会自动根据放入 res 目录中的 xml 界面配置文件、图片以及一些文本等资源文件同步更新修改 R.java 文件，所以不要随意手动编辑该文件。

assets/——资源目录，存放应用程序中用到的较大的文件，如视频文件、MP3 等一些媒体文件。注意：assets 目录下的资源文件不会在 R.java 自动生成 ID，所以读取 assets 目录下的文件必须指定文件的路径。

res/——资源文件夹，存放应用中用到的文字、图片以及布局等资源，包含程序中的所有文件。其中：

- res/drawable 用于存放图片文件资源。hdpi、mdpi、ldpi、xhdpi 和 xxhdpi 分别表示不同分辨率的图片。
- res/layout 用于存放显示界面（布局）的 xml 配置文件。
- res/menu 用于存放菜单设计 xml 文件。
- res/values 用于存放一些常量信息，如：strings.xml 用于定义字符串和数值、array.xml 用于定义数组信息、colors.xml 用于定义颜色、dimens.xml 用于定义尺寸数据、styles.xml 用于定义样式等。

android.jar——Android 程序应用的库文件，Android 支持的 API 都包含在这个文件夹里。

AndroidManifest.xml——项目清单文件，应用程序中的所有功能都在此列出。包括项目中的 Activity、Services、Broadcast Receiver 等信息，以及一些用户权限信息等。它是每个 Android 项目都必需的基础配置文件。

project.properties——项目环境信息，存放了项目的环境配置信息，一般不用编辑。

2. 几个重要的项目文件解析

（1）Java 源代码文件——MainActivity.java。当一个项目被创建后，会在 src 文件夹中生成一个源代码文件 MainActivity.java，其代码如下：

```
package com.app.hello;
Import android.os.Bundle;
Import android.app.Activity;
Import android.view.Menu;
Public class mainactivity extends Activity {
    @Override
    Protected void oncreate(Bundle savedinstancestate) {
        Super.oncreate(savedinstancestate);
        Setcontentview(R.layout.activity_main);
    }
    @Override
    Public boolean oncreateoptionsmenu(Menu menu) {
        Getmenuinflater().inflate(R.menu.main, menu);
        Return true;
    }
}
```

MainActivity 类继承自 Activity，Android 中所有的用户界面展示的类都直接或间接继承自 Activity。onCreate()是一个重要的方法，在这里，它重载 Activity 类中的 onCreate()方法，每个

Activity 类的子类都要重载该方法来初始化界面。R.layout.activity_main 是一个资源的常量，这个资源是对 activity_main.xml 的一个间接引用，当程序启动时将 activity_main.xml 文件中的内容展示给用户。

（2）Java 源代码文件——R.java。Android 应用程序 gen 目录下保存的是项目的所有包及源文件（.java），gen 目录下包含了项目中的所有资源。.java 文件是在建立项目时自动生成的，为只读模式，不能更改。R.java 文件是定义项目所有资源的索引文件。其部分代码如下：

```
package com.app.hello;
public final class R {
        public static final class attr {
        }
        public static final class dimen {
            public static final int activity_horizontal_margin
=0x7f040000;
            public static final int activity_vertical_margin
=0x7f040001;
        }
        public static final class drawable {
            public static final int ic_launcher=0x7f020000;
        }
        public static final class id {
            public static final int action_settings=0x7f080000;
        }
        public static final class layout {
            public static final int activity_main=0x7f030000;
        }
        ...
        }
}
```

以上代码定义了很多常量，它们的常量名都与 res 文件夹中的文件名相对应。有了这个文件，在程序中使用资源将更加方便，能很快地找到要使用的资源。当项目加入了新资源后，只需刷新一下该项目，.java 文件便会自动更新。

（3）资源文件——activity_main.xml。Android 中的 activity_main.xml 文件内容主要是有关用户界面布局和设计的，利用 XML 语言描述用户界面，为应用程序中使用的组件进行定义、设置界面属性值，以及组件使用的资源参考设置等。在 res/layout 目录下，双击 activity_main.xml 文件，其代码如下：

```
<RelativeLayout xmlns:android="http://schemas.android.com/apk/res/android"
      xmlns:tools="http://schemas.android.com/tools"
      android:layout_width="match_parent"
      android:layout_height="match_parent"
      android:paddingBottom="@dimen/activity_vertical_margin"
      android:paddingLeft="@dimen/activity_horizontal_margin"
      android:paddingRight="@dimen/activity_horizontal_margin"
      android:paddingTop="@dimen/activity_vertical_margin"
```

```
        tools:context=".MainActivity" >
    <TextView
        android:layout_width="wrap_content"
        android:layout_height="wrap_content"
        android:text="@string/hello_world" />
</RelativeLayout>
```

以上代码中大部分都是为组件设置布局参数，其格式为 android：属性=“属性值”。

（4）资源文件——strings.xml。Android 应用程序的 res 目录下有一个 value 子目录，其中的 strings.xml 文件是用来存放所有文本信息和数值。使用该文件有两个目的：一是为了程序国际化，可以将屏幕中可能出现的文字信息都集中存储在 strings.xml 文件中，当不同国家的用户使用时，只需修改 strings.xml 文件内容，而不用修改主程序。二是为了减少文字的重复使用，当一个提示信息需要在程序中使用多次时，就可以放在 strings.xml 文件中，需要时引用一下就可以。

strings.xml 文件代码如下：

```
<?xml version="1.0" encoding="utf-8"?>
<resources>
    <string name="app_name">HelloAndroid</string>
    <string name="action_settings">Settings</string>
    <string name="hello_world">Hello world!</string>
</resources>
```

代码中每个 string 标签声明了一个字符串，name 指定其引用名。value 文件中还有几个代表不同类别的 xml 文件，如 dimens.xml、styles.xml 它们的根元素都是<resources>，这样才可以识别调用资源。

（5）系统控制文件——AndroidManifest.xml。Android 包含四大组件，分别是：Activity、BroadCast receiver、service、Content Provider，当 Android 启动一个应用程序组件之前，必须知道哪些组件是存在的，所以开发人员在开发过程中，如果用到了这些组件，一定要在 AndroidManifest.xml 文件中声明，否则 Android 应用程序在运行时会报错。

这个文件以 XML 作为结构格式，而且对于所有应用程序，都叫做 AndroidManifest.xml。为声明一个应用程序组件，它还会做很多额外工作，比如指明应用程序所需链接到的库的名称（除了默认的 Android 库之外）以及声明应用程序期望获得的各种权限。当然 manifest 文件的主要功能还是向 Android 声明应用程序的组件。

以下是 AndroidManifest.xml 的代码实例：

```
<?xml version="1.0" encoding="utf-8"?>
<manifest xmlns:android="http://schemas.android.com/apk/res/android"
                                    //定义 Android 的命名空间
    package="com.app.hello"      //定义应用程序的包名称
    android:versionCode="1"
    android:versionName="1.0" >
    <uses-sdk
        android:minSdkVersion="8"
```

```
            android:targetSdkVersion="17" />
        <application
            android:allowBackup="true"
            android:icon="@drawable/ic_launcher"
            //定义应用程序的图标，资源类型为图像，名称为 ic_launcher
            android:label="@string/app_name" //定义应用程序的标签
            android:theme="@style/AppTheme"> //定义应用程序的主题？
            <activity                    //定义活动的内容
                android:name="com.app.hello.MainActivity"
    //定义活动的名称
                android:label="@string/app_name" >
                                        //定义 Android 应用程序的标签名称
                <intent-filter> //描述此 Activity 启动的位置和时间
                  <actionandroid:name=
    "android.intent.action.MAIN" />
                    <category android:name=
    "android.intent.category.LAUNCHER" />
                </intent-filter>
            </activity>
        </application>
    </manifest>
```

2.3.3 运行项目

上面我们已经利用 ADT 插件通过 Eclipse 创建好了第一个 Android 项目，而且没有编写任何代码，下面将其在模拟器上运行。

右击“HelloAndroid”项目名称，在菜单中选择“Run As”项目下的“Android Application”，便可以运行 helloandroid 项目了，不过 Android 模拟器启动非常慢，需要等待一会儿。启动后如图 2-27 所示。

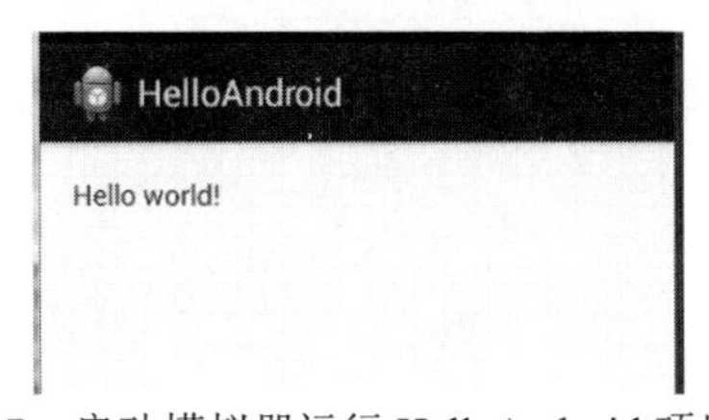

图 2-27　启动模拟器运行 HelloAndroid 项目界面

2.4 DDMS 应用

在完成了第一个“HelloAndroid”项目的创建之后，大家可以体会到在 ADT 开发环境中进行 Android 开发是一件很方便的事。实际上，ADT 还为用户提供了一个非常方便的调试工具，那就是 DDMS。使用这个工具可以将代码调试工作也变简单。DDMS 的全称是 Dalvik Debug Monitor Service，它为用户提供的服务包括：为测试设备截屏，针对特定的进程查看正在运行的线程以及堆信息、Logcat（用于得到程序 log 信息的命令）、广播状态信息、模拟电话呼叫、接收 SMS、虚拟地理坐标等。DDMS 为 Android 集成环境、模拟器及真正的 Android 设备架起了一座桥梁。

单击 Eclipse 界面 Window 下拉菜单中的“Open Perspective”，选择“DDMS”，切换到 DDMS 界面，如图 2-28、图 2-29 所示。单击图 2-29 右上角红色方框中的按钮，也可以切换界面。

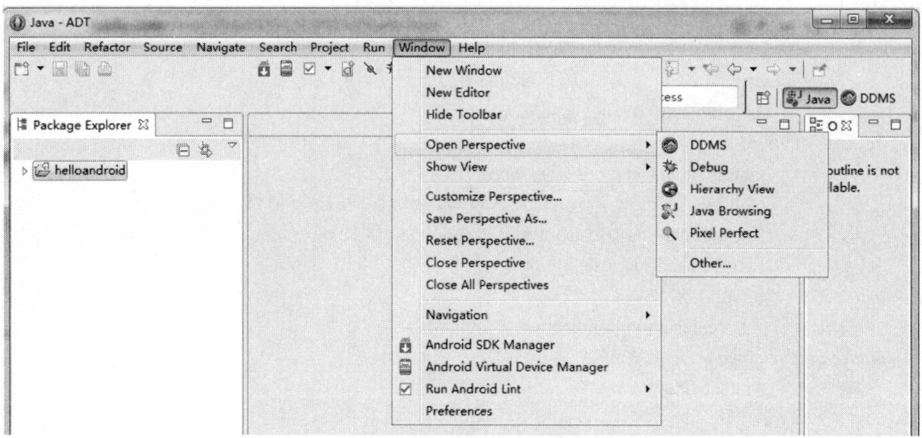

图 2-28　启动 DDMS

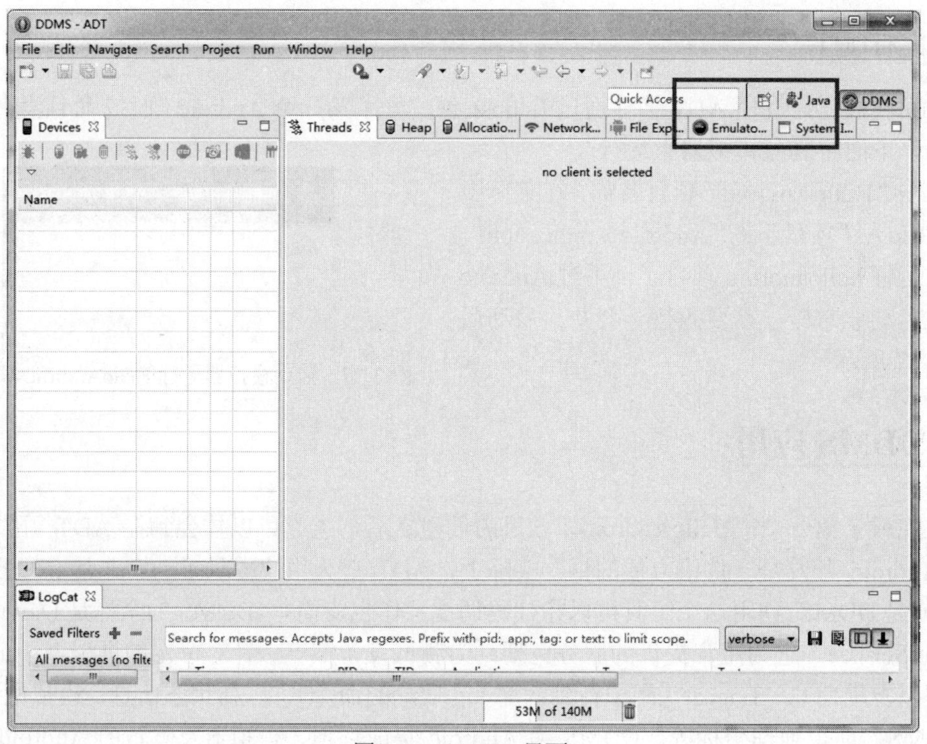

图 2-29　DDMS 界面

DDMS 布局中包含了四个主要部分：

（1）Devices。这个视图列出了当前连接到 Eclipse 的所有目标设备以及运行在目标设备上的进程。如果目标设备为手机模拟器，还会列出模拟器的端口号。Devices 视图的工具栏上还包含了一些按钮。比如屏幕截取按钮，它的外观是一个很小的 Android 手机屏幕，单击这个按钮以后，手机当前的屏幕就会被截取下来，开发者还可以将截取到的屏幕保存成 PNG 格式的图片以备它用。

（2）Threads/ Heap/ File Explorer。这三个视图可以让程序开发人员知道系统内部的运行状况。Threads 视图用来显示当前进程的所有活动线程。首先在 Devices 视图中选择想要查看的进程，然后单击 Devices 工具栏上的 Update Threads 按钮，该进程的所有活动线程信息就会显示在 Threads 视图中。如图 2-30 所示。

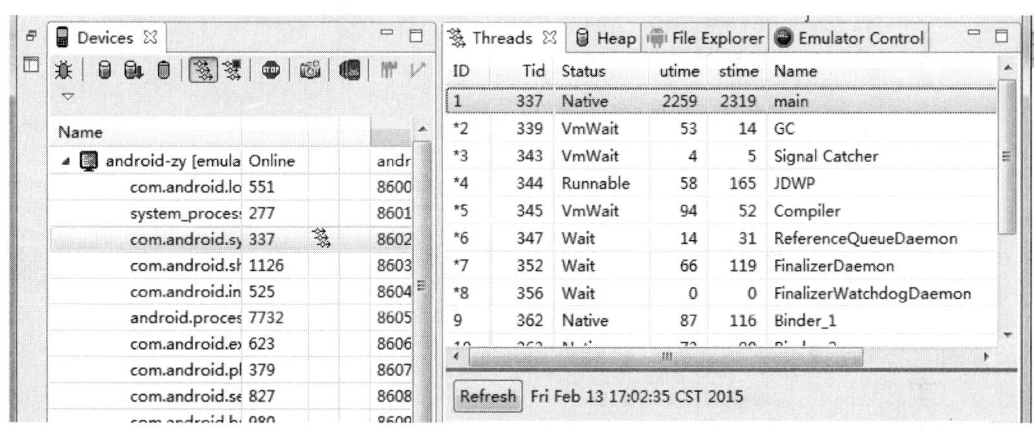

图 2-30　Thread 视图

Heap 视图用来显示 Dalvik 虚拟机中堆内存的使用情况，该视图的内容会在每次执行完垃圾回收操作之后进行更新。想要激活 Heap 视图必须单击 Devices 视图工具栏上的 Update Heap 按钮。系统的垃圾回收操作是不定时的，如果想要强制进行垃圾回收操作可以单击 Heap 视图中的 Cause GC 按钮。

File Explorer 视图是 Android 系统的文件浏览器，在这里可以浏览设备里面的文件目录，比如之前讲 Android 数据存储时提到过可以使用 DDMS 来浏览对应的存储文件，就是这个视图的功能。

（3）Emulator Control。这个视图仅适用于 Android 模拟器，用于模拟电话功能和定位功能：

1）模拟电话功能：用于模拟各种网络状态（离线、服务区内、漫游和搜寻网络等）和宽带下的语音通话、数据服务和短信息服务。这个功能对于测试应用程序的鲁棒性非常有用。

2）模拟来电或 SMS 文本消息：在“来电号码”字段中输入一个数字，然后单击呼叫，模拟呼叫发送到模拟器或手机。单击挂断键终止通话。在消息字段中输入文字，然后单击发送按钮发送消息。

3）模拟定位功能：用于设定手机模拟器的虚拟位置信息，输入模拟的经纬度。还可以指定一个 GPX 或者 KML 轨迹文件作为输入，用于模拟手机的移动状态，如图 2-31 所示。

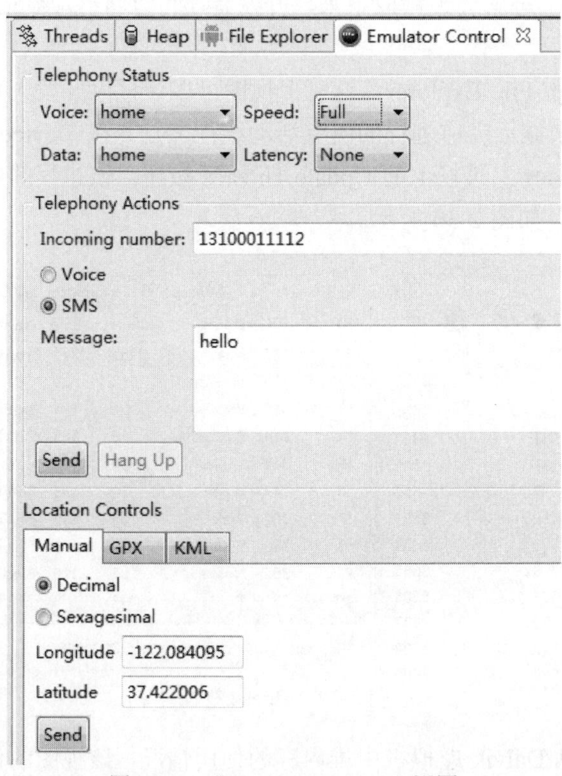

图 2-31 “Emulator Control”视图

（4）LogCat。这个视图用于打印设备的调试信息，在开发过程中最常用。这里的信息分为五级，分别对应上面的 V（VERBOSE）、D（DEBUG）、I（INFO）、W（WARN）、E（ERROR）五个圆形的按钮。此外，还可以通过单击这些按钮来过滤相应的调试信息。

2.5 Activity 介绍

2.5.1 Activity 的简介

Activity（活动）是 Android 组件中最基本也是最常用的一种组件。在 Android 系统中，Activity 是应用程序和用户交互的窗口，一个 Activity 通常就是一个单独的屏幕。Activity 是用户唯一可以看得到的东西。几乎所有的 Activity 都与用户进行交互，所以 Activity 主要负责的就是创建显示窗口，以便在其中存放各种显示控件，如菜单、文本输入等。Activity 展现在用户面前的经常是全屏窗口，也可以将 Activity 作为浮动窗口来使用，或者嵌入到其他的 Activity 中。一个 Android 程序由多个 Activity 程序组成，第一个呈现给用户的 Activity 称为 main Activity（主活动）。

通常称用户从一个屏幕界面转移到另一个屏幕界面的过程为 Activity 之间的切换。Activity 切换活动有两种类型：“独立”Activity 与“相依”Activity。它们的主要区别为是否与其他 Activity 交换信息。独立的 Activity 只是单纯从一个屏幕界面跳到下个界面，不涉及信息的交换。相依 Activity 又可分为单向与双向：从一个屏幕跳到下一个屏幕时，会把一些参数传给下一个 Activity 使用，这就是单向相依 Activity；要在两个屏幕之间切换，屏幕上的信息会因另一个屏幕的操作而改变的，就是双向相依 Activity。

2.5.2 创建 Activity

Activity 提供了和用户交互的可视化界面。将界面上的内容称为控件或者 View，用户通过对这些控件的操作，实现需求。Activity 提供了很多可直接使用的 View 控件，如：buttons（按钮）、menu items（菜单选项）、text fields（文本输入）、check boxes（选项）等。

1. 创建 Activity 的步骤

（1）一个 Activity 其实就是一个子类，这个类继承于 Activity 或其子类。

（2）覆盖 onCreate()方法，该方法在 Activity 第一次运行时，Activity 框架会调用这个方法。

（3）由于 Activity 是 Android 应用程序的一个组件，所以每一个 Activity 都需要在配置文件 AndroidManifest.xml 中进行配置。

（4）为 Activity 添加必要的控件。在 layout 文件夹中创建一个声明 xml 格式的布局文件，然后再在这个布局文件中对 Activity 的布局以及不同的控件进行设置。

（5）再在第一步定义的 Activity 子类中通过 findViewById(R 中对应的 id 类中控件的 id)方法来获取布局文件中声明的控件，前提是布局文件 R 中必须声明这些控件的 id。

2. 创建实例

（1）右击 com.example.lp 包→【New】，选择新建一个 Class 类文件，如图 2-32 所示。

（2）创建一个名为 firstActivity 的类，单击【Finish】完成创建，如图 2-33 所示。

（3）重写 Activity 的 onCreate()方法，代码如下：

```
public class firstActivity extends Activity {
        @Override
    protected void onCreate(Bundle savedInstanceState) {
        super.onCreate(savedInstanceState);//调用了父类的 onCreate()方法
    }
}
```

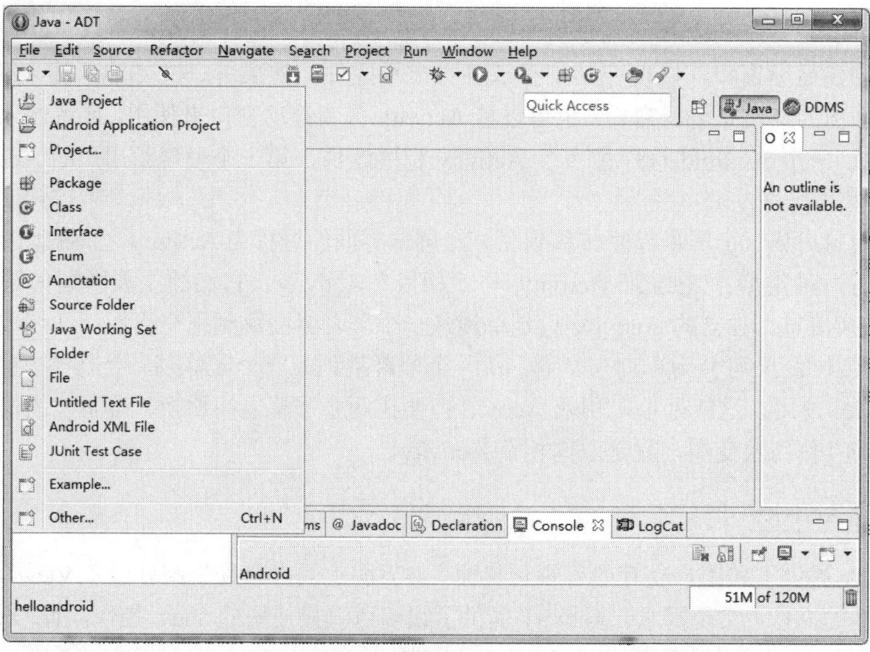

图 2-32　新建一个 Class

图 2-33　新建一个 firstActivity 类

2.6 本章小结

本章首先讲解了 Android 开发环境的搭建。要进行 Android 程序开发，需要下载安装 JDK 以及 ADT，并完成环境变量的配置即可。然后，介绍了 Android 开发工具的界面及创建启动 AVD 的过程。接着，讲解了在 Android 系统中建立并测试一个程序项目的过程，分析了应用程序项目的架构和重要文件。此外，还简单介绍了 ADT 为用户提供的便捷调试工具 DDMS 的使用方法。最后，介绍了 Android 系统四大组件之一的 Activity 的主要知识和创建方法。通过本章的学习，对 Android 程序设计流程有一个全面的认识。

2.7 本章习题

1．请简要说明 Android 应用中有哪些主要组件，并描述其作用。
2．创建 Android 应用，显示“Android ACTIVITY 组件”文本信息。

3

Android 用户界面开发

界面通常包括图形和文字，对于 Android 应用开发人员来说，了解用户界面设计方法非常有必要。一个好的应用界面应包括内容清楚、屏幕美观并具有亲和力和交互性强等优点。应用界面的设计需要反复推敲和分析，达到用户的满意是终极目标，Android 平台提供了非常完美的控件供开发者设计。本章将详细介绍 Android 用户界面的开发方法，供开发者学习。

学习目标：

了解 Android 中用户界面开发设计原则及方法，掌握常用界面组件的设计和使用。

【知识目标】

- 掌握界面布局设计方法
- 掌握常用界面组件
- 掌握界面资源的定义与使用
- 掌握单选按钮、多项选择和相关事件

【技能目标】

- 掌握 Android 用户界面开发方法
- 能熟练运用所学的知识点，完成开发标准身高计算器、制作手机桌面、调查问卷等实训项目

3.1 用户界面开发详解

用户界面是系统和用户之间进行交互和信息交换的媒介，它实现了信息的内部形式与人类可以接受的形式之间的转换。下面我们从用户界面设计原则和用户界面设计核心概念这两个方面对用户界面进行详细的介绍。

3.1.1 用户界面设计原则

为了能够更好地进行界面设计，避免一些不必要的问题的出现，对于用户界面的设计通常有一些原则需要遵守。

1. 清晰简洁的视觉层次

设计时要让用户把注意力放在最重要的地方。单列布局对于全局会有更好的掌控，过多的列有分散用户注意力的风险，使用户的主旨无法很好表达，清晰的层级关系也将对降低外观的复杂性起到重要作用。

2. 界面设计要保持一致性

在设计中保持一致性可以减少用户的学习成本，用户不需要学习新的操作，用户进行点击或拖拽所得到的结果在整个程序的各个界面都应是一致的。除此之外，包括颜色、方向、元素的表现形式，位置、大小、形状等方面也应表现出一致性，一致性的界面可以让用户对于如何操作有更好的理解，从而提升效率。

3. 增强容错能力并放宽输入要求

对用户输入的数据，尽量放宽限制，包括格式、大小写。如果用户的行为引起了一个错误，在恰当的时机运用信息显示什么行为是错误的，并确保用户明白如何防止这种错误的再次发生，尽可能不要让其重新填写全部内容。

4. 了解用户并给出产品定位

用户是终极评判人，因此我们必须了解用户，通过不断了解目标客户的需求及标准，才能把产品做得更好，得到更多与客户交流的机会。不要迷恋于追逐设计趋势的更新，或者是不断添加新的功能。始终记住，首要的任务是关注用户，这样才能创造出一个能让用户达成目标的界面。

3.1.2 用户界面设计核心概念

用户界面设计是指对软件的人机交互、操作逻辑、界面美观的整体设计。设计的核心目的就是要让用户拥有更好的使用体验，让软件的操作变得舒适、简单、自由，充分体现软件的定位和特点。

通过不同的界面设计，来展现不同的设计元素及设计方式对用户行为的影响，以此来实现多元化、个性化的用户要求，及具有交互性的网络界面设计目标。

3.2 界面布局设计

Android 包含 LinearLayout、RelativeLayout、TableLayout、AbsoluteLayout 和 FrameLayout 等多种布局。有两种实现布局的方式：一种是通过布局参数类提供的方法设置，另一种是通过 XML 属性设置。

（1）LinearLayout。LinearLayout 是一种线性的布局，在该布局中子元素之间呈线性排列，即顺序排列。由于布局是显示在二维空间中的，因此顺序排列是指在某一方向上的顺序排列，常见的有水平顺序排列和垂直顺序排列。

（2）RelativeLayout。RelativeLayout 是一种根据相对位置排列元素的布局，这种方式允许子元素指定它们相对于其他元素的位置（通过 ID 指定）。相对于线性布局，子元素可任意放置，没有规律性。需要注意，线性布局不需要特殊指定其父元素，但在使用之前必须指定其参照物，只有指定参照物之后，才能定义其相对位置。

（3）TableLayout。同 LinearLayout 类似，TableLayout 是一种表格布局，这种布局将子元素的位置分配到行或列中，即按照表格的顺序排列。一个表格布局有多个"表格行，而每个表格行又包含表格单元。需要注意，表格布局并不是真正意义上的表格，只是按照表格的方式组织子元素的位置。在表格布局之中，子元素之间并没有实际表格中的分隔线。

（4）AbsoluteLayout。相对布局需要指定其参照的父元素，AbsoluteLayout（绝对布局）与相对布局相反，绝对布局不需要指定其参照物，绝对布局使用整个手机界面作为坐标系，通过坐标系的两个偏移量（水平偏移量和垂直偏移量）来唯一指定其位置。

（5）FrameLayout。FrameLayout 称为帧布局或层布局，将组件显示在屏幕的左上角，后面的组件覆盖前面的组件，除非后面的组件是透明的，否则前面的组件将不可见。

3.2.1 线性布局

线性布局分为垂直线性布局和水平线性布局，垂直线性布局是通过 android:orientation=“vertical”来实现，水平线性布局是通过 android:orientation=“horizontal”来实现。线性布局运行效果如图 3-1 所示，LinearLayout 常用属性及对应设置方法见表 3-1，gravity 可取的属性及说明表见表 3-2。

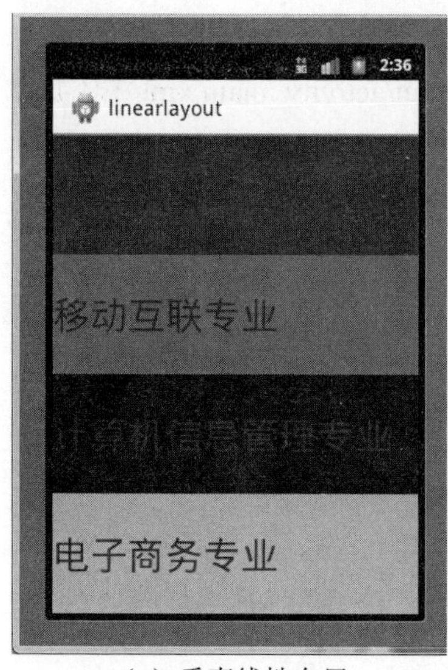

（a）垂直线性布局

（b）水平线性布局

图 3-1　线性布局

表 3-1　LinearLayout 常用属性及对应设置方法

属性名称	对应方法	描述
android:orientation	setOrientatin(int)	设置线性布局的朝向，可取 Horizontal(水平)和vertical（垂直）
android：gravity	setGravity(int)	设置线性布局的内部元素的布局方式

表 3-2　gravity 可取的属性及说明表

属性值	说明
top	不改变控件的大小，对齐到容器顶部
bottom	不改变控件的大小，对齐到容器底部
left	不改变控件的大小，对齐到容器左部
right	不改变控件的大小，对齐到容器右部
center_vertical	不改变控件的大小，对齐到容器纵向中央位置
center_horizontal	不改变控件的大小，对齐到容器横向中央位置
center	不改变控件的大小，对齐到容器中央位置
fill_vertical	若有可能，纵向拉伸以填充容器
fill_horizontal	若有可能，横向拉伸以填充容器
fill	若有可能，纵向、横向同时拉伸以填充容器

【实例 3.1】垂直线性布局实例［完成图 3-1（a）的线性布局方式］。

步骤：创建名为“LinearLayout”的工程，在 res/layout/activity_main.xml 中添加如下代码：

```
<LinearLayout
        android:layout_width="fill_parent"//设置布局的宽度填充整个界面
        android:layout_height="fill_parent"
        android:orientation="vertical" >//设置线性布局的朝向
    <TextView
        android:id="@+id/textView1"
        android:layout_width="fill_parent"
        android:layout_height="wrap_content"//设置布局高度随内容调整，以显示全部的内容
        android:text="计算机应用专业"
        android:gravity="center_vertical"
        android:textSize="15pt"//设置布局的文字字号
        android:background="#aa0000"//设置布局的背景颜色
        android:layout_weight="1" />//设置布局的权值，按比例分配界面
    <TextView
        android:id="@+id/textView2"
        android:layout_width="fill_parent"
        android:layout_height="wrap_content"
        android:text="移动互联专业"
        android:gravity="center_vertical"
        android:textSize="15pt"
        android:background="#00bb00"
        android:layout_weight="1"    />
    <TextView
        android:id="@+id/textView3"
        android:layout_width="fill_parent"
        android:layout_height="wrap_content"
        android:text="计算机信息管理专业"
        android:gravity="center_vertical"
        android:textSize="15pt"
        android:background="#0000cc"
        android:layout_weight="1" />
    <TextView
        android:id="@+id/textView4"
        android:layout_width="fill_parent"
        android:layout_height="wrap_content"
        android:text="电子商务专业"
        android:gravity="center_vertical"
        android:textSize="15pt"
        android:background="#aabbcc"
        android:layout_weight="1"    />
</LinearLayout>
```

【实例 3.2】水平线性布局实例［完成图 3-1（b）的线性布局方式］。

步骤：创建名为“horizontalExample”的工程，在 res/layout/activity_main.xml 中添加如下代码：

```
<LinearLayout
    android:layout_width="fill_parent"
    android:layout_height="fill_parent"
    android:orientation="horizontal" >
    <TextView
        android:id="@+id/textView1"
        android:layout_width="wrap_content"
        android:layout_height="fill_parent"
        android:text="计算机应用专业"
        android:gravity="center_horizontal"
        android:textSize="4pt"
        android:background="#aa0000"
        android:layout_weight="1" />
    <TextView
        android:id="@+id/textView2"
        android:layout_width="wrap_content"
        android:layout_height="fill_parent"
        android:text="移动互联专业"
        android:gravity="center_horizontal"
        android:textSize="4pt"
        android:background="#00bb00"
        android:layout_weight="1"   />
    <TextView
        android:id="@+id/textView3"
        android:layout_width="wrap_content"
        android:layout_height="fill_parent"
        android:text="计算机信息管理专业"
        android:gravity="center_horizontal"
        android:textSize="4pt"
        android:background="#0000cc"
        android:layout_weight="1" />
    <TextView
        android:id="@+id/textView4"
        android:layout_width="wrap_content"
        android:layout_height="fill_parent"
        android:text="电子商务专业"
        android:gravity="center_horizontal"
        android:textSize="4pt"
        android:background="#aabbcc"
        android:layout_weight="1"   />
  </LinearLayout>
```

3.2.2 相对布局

线性布局有时不能满足我们的需要，例如我们要在一行（列）上显示多个控件，这就需要使用 Relativelayout 来进行相对布局，Relativelayout 允许子元素指定它们相对于其他元素或父元素的位置（通过 ID 指定）。因此，我们可以右对齐、上下或置于屏幕中央的形式来排列两个元素。Relativelayout 视图显示了屏幕元素的类名称，表 3-3 是每个元素的属性列表。这些属性一部分由元素直接提供，另一部分由容器的 Layoutparams 成员（Relativelayout 子类）提供。

表 3-3　相对布局中属性值只取 true 或 false 的属性表

属性名称	属性说明
android:layout_centerHorizontal	当前控件位于父控件的横向中间位置
android:layout_centerVertical	当前控件位于父控件的纵向中间位置
android:layout_centerInParent	当前控件位于父控件的中间位置
android:layout_alignParentBottom	当前控件底部与父控件的底部对齐
android:layout_alignParentLeft	当前控件左侧与父控件的左侧对齐
android:layout_alignParentRight	当前控件右侧与父控件的右侧对齐
android:layout_alignParentTop	当前控件顶部与父控件的顶部对齐
android:layout_alignWithParentMissing	参照控件不存在或不可见时参照父控件

Relativelayout 参数有 Width、Height、Below、AlignTop、ToLeft、Padding 和 MarginLeft。注意，这些参数中的一部分，其值是相对于其他子元素而言的，所以才被称为 Relativelayout 布局。这些参数包括 ToLeft、AlignTop 和 Below，用于指定相对于其他元素的左、上和下的位置见表 3-4，相对布局中取值为像素的属性及说明表见表 3-5。

表 3-4　相对布局中属性值为其他控件 id 属性及说明表

属性名称	属性说明
android:layout_toRightOf	使当前控件位于给出 id 控件的右侧
android:layout_toLeftOf	使当前控件位于给出 id 控件的左侧
android:layout_above	使当前控件位于给出 id 控件的上方
android:layout_below	使当前控件位于给出 id 控件的下方
android:layout_alignTop	使当前控件的上边界与给出 id 控件的上边界对齐
android:layout_alignBottom	使当前控件的下边界与给出 id 控件的下边界对齐
android:layout_alignLeft	使当前控件的左边界与给出 id 控件的左边界对齐
android:layout_alignRight	使当前控件的右边界与给出 id 控件的右边界对齐

表 3-5　相对布局中取值为像素的属性及说明表

属性名称	属性说明
android：layout_marginLeft	当前控件左侧留白
android：layout_marginRight	当前控件右侧留白
android：layout_marginTop	当前控件顶部留白
android：layout_marginBottom	当前控件底部留白

注意：在进行相对布局时要避免出现循环依赖，例如设置相对布局在父容器中的排列方式为 WRAP_CONTENT，就不能再将相对布局的子控件设置为 ALIGN_RARENT_BOTTOM。因为这样会造成子控件和父控件相互依赖和参照错误。

【实例 3.3】设计一个带按钮的输入框，完成相对布局实例（完成如图 3-2 所示的相对布局方式）。

图 3-2　相对布局方式

步骤：创建名为“RelativElayout”的工程，在 res/layout/activity_main.xml 中添加如下代码：

```
<RelativeLayout
        android:layout_width="fill_parent"
        android:layout_height="fill_parent" >
         <TextView
                android:id="@+id/label1"
                android:layout_width="fill_parent"
                android:layout_height="wrap_content"
                android:text="请您输入用户名：" />
          <EditText
                android:id="@+id/enter"
                android:layout_width="fill_parent"
                android:layout_height="wrap_content"
                android:background="@android:drawable/editbox_background"
                android:layout_below="@id/label1" >
                //将该 EditText 控件的底部置于给定 TextView ID 的控件之下;
          </EditText>
          <Button
                android:id="@+id/ok"
                android:layout_width="wrap_content"
                android:layout_height="wrap_content"
                android:layout_below="@+id/enter"
                //将该 Button 控件的底部置于给定 TEditText 控件之下
                android:layout_alignParentRight="true"
```

```
                //将该控件的右部与其父控件 Edittext 控件的右部对齐
                android:layout_marginLeft="10dip"
                //左偏移的值为 10dip
                android:text="确定" />
            <Button
                android:id="@+id/button2"
                android:layout_width="wrap_content"
                android:layout_height="wrap_content"
                android:layout_toLeftOf="@id/ok"
                android:layout_alignTop="@+id/ok"
                android:text="取消" />
</RelativeLayout>
```

3.2.3 绝对布局

绝对布局是指屏幕中所有控件的摆放都由开发人员通过设置控件的坐标来指定，控件容器不再负责管理其子控件的位置。由于子控件的位置和布局都通过坐标来指定，因此 AbsoluteLayout 类中并没有开发特有的属性和方法。

【实例 3.4】设计一个登录系统界面完成绝对布局实例（完成如图 3-3 所示的绝对布局方式）。

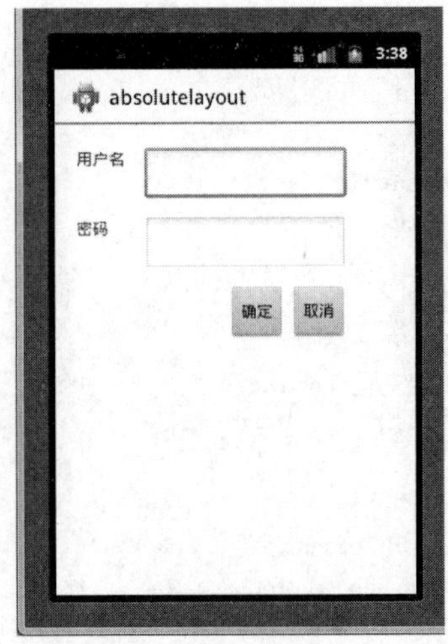

图 3-3 绝对布局方式

步骤：创建名为“AbsoluteLayout”的工程，在 res/layout/activity_main.xml 中添加如下代码：

```
<AbsoluteLayout
        android:id="@+id/AbsoluteLayout01"
        android:layout_width="fill_parent"
        android:layout_height="fill_parent">
        <TextView
```

```
        android:id="@+id/textView1"
        android:layout_x="20dip"
        android:layout_y="20dip"
        android:layout_width="wrap_content"
        android:layout_height="wrap_content"
        android:text="用户名" />
    <TextView
        android:id="@+id/textView2"
        android:layout_width="wrap_content"
        android:layout_height="wrap_content"
        android:layout_x="20dip"
        android:layout_y="80dip"
        android:text="密码" />
    <EditText
        android:id="@+id/editText2"
        android:layout_width="180dip"
        android:layout_height="wrap_content"
        android:layout_x="80dip"
        android:layout_y="80dip"
        android:password="true">
      </EditText>
      <EditText
        android:id="@+id/editText1"
        android:layout_width="180dip"
        android:layout_height="wrap_content"
        android:layout_x="80dip"
        android:layout_y="20dip"
        android:ems="10" />
    <Button
        android:id="@+id/button1"
        android:layout_width="wrap_content"
        android:layout_height="wrap_content"
        android:layout_x="155dip"
        android:layout_y="140dip"
        android:text="确定" />
    <Button
        android:id="@+id/button2"
        android:layout_width="wrap_content"
        android:layout_height="wrap_content"
        android:layout_x="210dip"
        android:layout_y="140dip"
        android:text="取消" />
</AbsoluteLayout>
```

3.2.4 表格布局

TableLayout 将子元素的位置分配到行和列中。Android 的一个 TableLayout 由许多的

TableRow 组成，每个 TableRow 都会定义一个 Row。TableLayout 容器不会显示 Row、Column 或 Cell 的边框线。每个 Row 拥有 0 个或多个 Cell，每个 Cell 拥有一个 View 对象。表格由行和列组成许多单元格，允许单元格为空。单元格不能跨列，这与 HTML 中不一样。列可以被隐藏，也可以被设置为伸展的，从而填充可利用的屏幕空间，也可以被设置为强制列收缩，直到表格匹配屏幕大小。具体常用属性及其对应方法说明见表 3-6，表格布局实例如图 3-4 所示。

表 3-6　TableLayout 类常用属性及其对应方法说明表

属性名称	对应方法	描述
android：collapseColumns	setCoiumnCollapsed(int,boolean)	设置指定列号的列为 Collapsed，列号从 0 开始计算
android：shrinkColumns	setShrinkCollumns(boolean)	设置指定列号的列为 Shrinkable，列号从 0 开始计算
android：stretchColumns	setStretchAllColumns(boolean)	设置指定列号的列为 Stretchable，列号从 0 开始计算

例如：

```
<TableLayout android:id="@+id/TableLayout01"
android:layout_width="fill_parent"
android:layout_heigth="wrap_content"
android:background="@color/white"
android:collapseColumns="1"
//隐藏编号为 1 的列，若有多个列要隐藏，则用逗号隔开，如 0,2
android:shrinkColumns="0"
//设置 0 号列为可收缩的列，可收缩的列会纵向扩展，若有多个列要收缩，则用逗号隔开，如 0,2>
```

【实例 3.5】表格布局实例（完成图 3-4 的表格布局方式）。

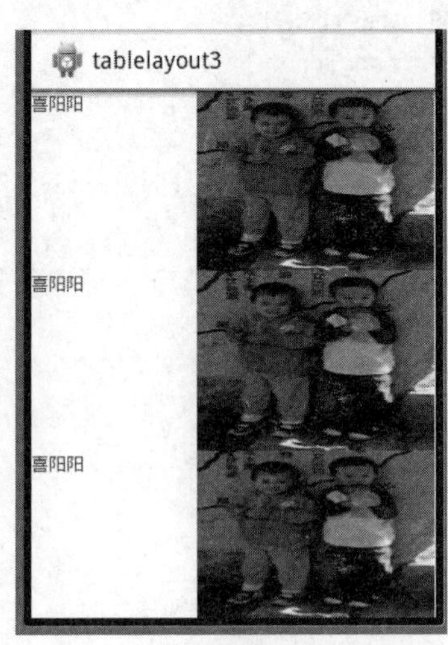

图 3-4　表格布局

步骤一：创建名为“TableLayout”的工程，在 res/layout/activity_main.xml 中添加如下代码：

```
<?xml version="1.0" encoding="utf-8"?>
<LinearLayout xmlns:android="http://schemas.android.com/apk/res/android"
        android:layout_width="fill_parent"
        android:layout_height="fill_parent"
        android:orientation="vertical"
        android:gravity="top">
            <TableLayout
                android:id="@+id/lyy"
                android:layout_width="fill_parent"
                android:layout_height="wrap_content">
            </TableLayout>
    </LinearLayout>
```

步骤二：在 src/MainActivity.java 中，添加如下代码：

```
public class MainActivity extends ActionBarActivity {
@Override
public void onCreate(Bundle bundle){
try
{
  super.onCreate(bundle);//必须执行父类的 onCreate 方法
  setContentView(R.layout.activity_main); //使用 activity_main 初始化程序布局
  TableLayout tableLayout=(TableLayout)findViewById(R.id.lyy); //创建 TableLayout 对象
  tableLayout.setStretchAllColumns(true);
  for(int row=0;row<3;row++)
  {
      TableRow tableRow=new TableRow(this);
      TextView tv=new TextView(this);
      ImageView iv=new ImageView(this);
      tv.setText("喜阳阳");
      iv.setBackgroundResource(R.drawable.pic);
      tableRow.addView(tv); //表格行添加文本框视图对象
      tableRow.addView(iv); //表格行添加图片视图对象
      tableLayout.addView(tableRow,new TableLayout.LayoutParams(ViewGroup.LayoutParams.FILL_
      PARENT, ViewGroup.LayoutParams.WRAP_CONTENT));
      //新建的 tablerow 添加到 TableLayout 布局，第一个参数为宽的设置，第二个参数为高的设置
      }//循环添加 3 个表格布局列
  }
  catch(Exception e){
  Toast.makeText(MainActivity.this,"异常错误："+e.toString(),Toast.LENGTH_LONG).show();
  }//catch 捕捉异常，若出现异常，则显示异常的 Toast 消息
  finally
  {}//添加处理异常代码
}
}
```

3 Chapter

3.2.5 帧布局

FrameLayout 帧布局在屏幕上开辟出了一块区域，在这块区域中可以添加多个子控件，但是所有的子控件都被对齐到屏幕的左上角。帧布局的大小由子控件尺寸最大的子控件来决定。在 FrameLayout 中，子控件是通过栈来绘制的，所以后添加的子控件会被绘制在上层。如果子控件一样大，同一时刻只能看到最上面的子控件。帧布局实例如图 3-5 所示。

帧布局 FrameLayout 继承自 ViewGroup，除了继承自父类的属性和方法，Framelayout 类中包含了自己特有的属性和方法，见表 3-7。

表 3-7 帧布局属性及对应方法表

属性	对应方法	描述
android:foreground	setforeground(Drawable)	设置绘制在所有子控件之上的内容
android: foregroundGravity	foregroundGravity(Drawable)	设置绘制在所有子控件之上的内容的 gravity 属性

【实例 3.6】帧布局实例（完成图的 3-5 帧布局方式）。

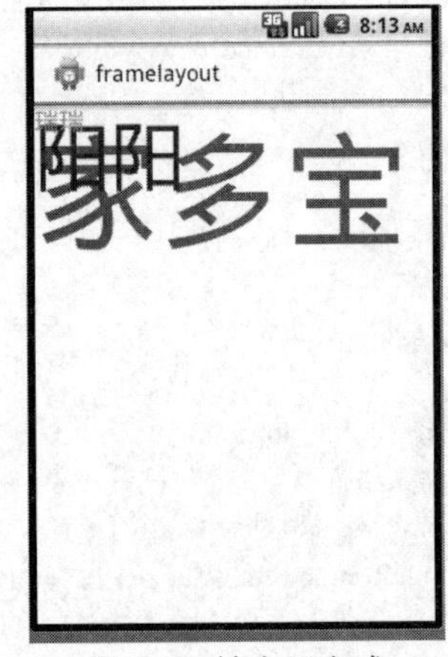

图 3-5 帧布局方式

步骤：创建名为“FrameLayout”的工程，在 res/layout/activity_main.xml 中添加如下代码：

```
<FrameLayout xmlns:android="http://schemas.android.com/apk/res/android"
    xmlns:tools="http://schemas.android.com/tools"
    android:id="@+id/container"
    android:layout_width="match_parent"
    android:layout_height="match_parent"
    tools:context="com.example.framelayout.MainActivity"
    tools:ignore="MergeRootFrame" >
    <TextView
```

```
            android:id="@+id/textView1"
            android:layout_width="wrap_content"
            android:layout_height="wrap_content"
            android:text="家多宝"
            android:textSize="100px"
            android:textColor="#ff0000"/>
        <TextView
            android:id="@+id/textView2"
            android:layout_width="wrap_content"
            android:layout_height="wrap_content"
            android:text="阳阳"
             android:textSize="60px"
             android:textColor="#0000ff"/>
        <TextView
            android:id="@+id/textView3"
            android:layout_width="wrap_content"
            android:layout_height="wrap_content"
            android:text="瑞瑞"
            android:textSize="20px"
            android:textColor="#00ff00" />
    </FrameLayout>
```

3.3 常用界面组件

Android 平台提供了大量组件，可以轻松构建 GUI 应用。本节将详细学习 Android 平台的各种组件的创建、属性，以及组件常用的事件处理及相关实例。

3.3.1 文本组件

TextView 是最见的一种组件，用来向用户显示文本，如邮件正文或应用程序标签等。API 中对应的类为 android.widget.TextView。布局文件里的一些常用的 xml 属性见表 3-8。

表 3-8 布局文件里的一些常用的 XML 属性

属性	解释
android:gravity	用来设置控件内文本的对齐方式
android:android_gravity	相对于父控件来说，用于设置控件的对齐方式
android:text	用来设置控件文本信息
android:layout_width	用来设置控件的宽度
android:layout_height	用来设置控件的高度
android:background	用来设置控件的背景颜色
android:textColor	用来设置控件的内文本的颜色
Android:textSize	用来设置控件的内文本字体大小

【实例 3.7】TextView 的应用，请设计一个如图 3-6 所示的界面。

图 3-6　TextView 组件设计界面

步骤一：创建名为“TextViewExample”的工程，在 res/layout/activity_main.xml 中添加如下代码：

```
<LinearLayout xmlns:android="http://schemas.android.">
      android:layout_width="fill_parent"
      android:layout_height="fill_parent"
      android:orientation="vertical"
    <TextView
         android:id="@+id/editText1"
         android:layout_width="fill_parent"
         android:layout_height="wrap_content"
         android:ems="6"
         android:text="天津电子信息职业技术学院" />
</LinearLayout>
```

Linearlayout 表示该 Activity 为线性布局，其中 layout_width 和 layout_height 分别定义长度和宽度，fill_parent 表示布满整个布局，wrap_content 表示根据内容动态布局，orientation 参数用于控制布局方向，vertical 表示垂直方向，horizontal 表示水平布局。TextView 控件中的 text 参数表示 TextView 要显示的文本，这里的文本值为 res/value/strings.xml 中定义的 text 元素的值。

步骤二：在 src 目录下打开 MainActivity.java 文件，修改文件，内容如下：

```
public class MainActivity extends ActionBarActivity {
@Override
      protected void onCreate(Bundle savedInstanceState)
      {
            super.onCreate(savedInstanceState);
            setContentView(R.layout.activity_main);
```

```
                TextView textView=(TextView)findViewById(R.id.editText1);
                textView.setTextColor(Color.RED);//设置文本颜色
                textView.setTextSize(30);//设置文本字号大小
                textView.setBackgroundColor(Color.BLUE);
                //设置文本背景颜色
        }
}
```

步骤三：运行项目。

3.3.2 按钮组件及相关的事件

按钮是使用最多的控件，在 Android 平台中，按钮是通过 Button 来实现的，实现过程也非常简单。既然是按钮，被点击之后必定要触发事件，所以需要对按钮设置 setOnClickListener、setOnLongClickListener 和 setOnTouchListener 等相关事件监听，下面我们分别介绍这三个触发事件。

一、Button 和 setOnClicklistener 单击事件

对于一个 Android 应用程序来说，事件处理是必不可少的，用户与应用程序之间的交互是通过事件处理来完成的。当用户与应用程序交互时，一定是通过触发某些事件来完成的，让事件来通知程序应该执行哪些操作，在这个繁杂的过程中主要涉及两个对象，事件源与事件监听器。事件源指的是事件所发生的控件，各个控件在不同情况下触发的事件不尽相同，而且产生的事件对象也可能不同。监听器则是用来处理事件的对象，实现了特定的接口，根据事件的不同重写不同的事件处理方法来处理事件。将事件源与事件监听器联系到一起，就需要为事件源注册监听，当事件发生时，系统才会自动通知事件监听器来处理相应的事件。如图 3-7 所示说明了事件处理的整个流程。

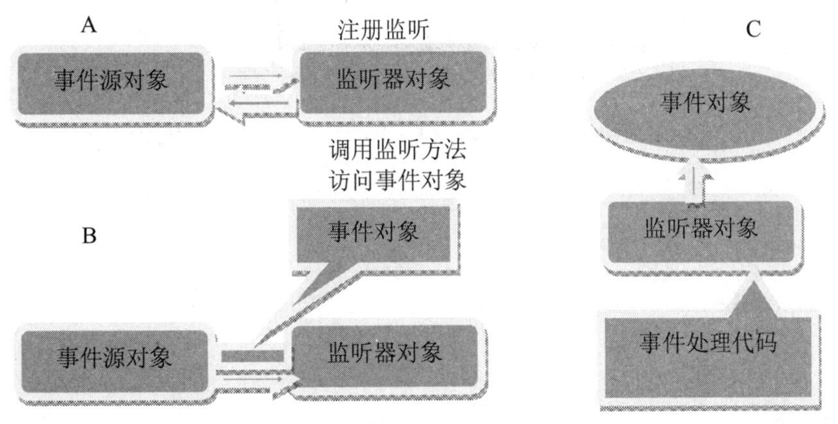

图 3-7 事件处理流程

事件处理的过程一般分为三步，如下所示：

第一步：为事件源对象添加监听，这样当某个事件被触发时，系统才会知道通知谁来处理该事件，如事件处理流程图 3-7（A）。

第二步：当事件发生时，系统会将事件封装成相应类型的事件对象，并发送给注册到事件源的事件监听器，如事件处理流程图 3-7（B）

第三步：当监听器对象收到事件对象后，系统会调用监听器中相应的事件处理方法来处理事件并给出响应，如事件处理流程图 3-7（C）。

【实例 3.8】Button 按钮和 onclicklistener 事件完成程序设计，如图 3-8 所示。当单击“移动互联专业”按钮时，立即在按钮上方显示文本“您选择了移动互联专业，欢迎你！”。

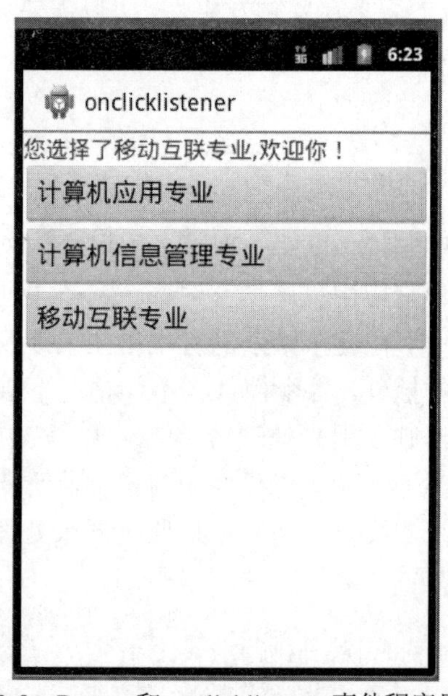

图 3-8　Button 和 onclicklistener 事件程序界面

步骤一：创建名为“onclicklistenerExample”的工程，在 res/layout/activity_main.xml 中添加如下代码：

```
<LinearLayout
        android:layout_width="match_parent"
        android:layout_height="match_parent"
        android:orientation="vertical">
        <TextView
            android:id="@+id/textView1"
            android:layout_width="wrap_content"
            android:layout_height="wrap_content"
            android:text="请您选择系部：" />
        <Button
            android:id="@+id/button1"
```

```
            android:layout_width="fill_parent"
            android:layout_height="wrap_content"
            android:text="计算机应用专业"
            android:gravity="left"
            android:textSize="20px" />
        <Button
            android:id="@+id/button2"
            android:layout_width="fill_parent"
            android:layout_height="wrap_content"
            android:text="计算机信息管理专业"
            android:gravity="left"
            android:textSize="20px"
            />
        <Button
            android:id="@+id/button3"
            android:layout_width="fill_parent"
            android:layout_height="wrap_content"
            android:text="移动互联专业"
            android:gravity="left"
            android:textSize="20px" />
</LinearLayout>
```

步骤二：在 src 目录下打开 MainActivity.java 文件，修改文件，内容如下：

```
public class MainActivity extends ActionBarActivity implements OnClickListener {
        Button[] buttons=new Button[3];
            TextView textView;
        public void onCreate(Bundle savedInstanceState){
            super.onCreate(savedInstanceState);
            setContentView(R.layout.activity_main);
          buttons[0]=(Button)this.findViewById(R.id.button1);
          buttons[1]=(Button)this.findViewById(R.id.button2);
          buttons[2]=(Button)this.findViewById(R.id.button3);
          textView=(TextView)this.findViewById(R.id.textView1);
          textView.setTextSize(18);
          for(Button button:buttons){
              button.setOnClickListener(this);}}
        @Override
        public void onClick(View v){
        textView.setText("您选择了"+((Button)v).getText()+",欢迎你！");
        }
}
```

步骤三：运行项目。

二、Button 和 OnLongClickListener 长按事件

对于一个 Android 应用程序来说，事件处理是必不可少的，用户与应用程序之间的交互便是通过事件处理来完成的。当用户与应用程序交互时，一定是通过触发某些事件来完成的，让

事件来通知程序应该执行哪些操作，在这个繁杂的过程中主要涉及两个对象，事件源与事件监听器 OnLongClickListen 接口与 OnClickListener 接口原理基本相同，只是该接口为 View 长按事件的捕捉接口，当长时间按下某个 View 时触发的事件，该接口对应的回调方法为：

```
public boolean onLongClick(View v)
```

参数 v：事件源控件，当长时间按下此控件时才会触发该方法。

返回值：该方法的返回值为 boolean 类型的变量，当返回 true 时，表示已经完整处理了这个事件，并不希望其他的调回方法再次进行处理；当返回 false 时，表示并没有完全处理完该事件，希望其他方法继续对其进行处理。

我们只需要修改上例例题里面的三个地方：

（1）定义类的接口为 OnLongClickListener。

（2）绑定监听器为：buttonsetOnLongClicklistener（this）。

（3）监听器改为

```
public boolean onLongClick(View v){
        textView.setText("您选择了"+((Button)v).getText()+",欢迎你！");
        return true;
        }
```

【实例 3.9】Button 和 onLongClicklistener 事件完成程序设计，如图 3-9 所示。

（a）项目启动后的界面

（b）按钮被按下后的界面

（c）鼠标拖动按钮的状态

图 3-9　ontouchlistener 程序设计界面

步骤一： 创建名为“onlongclicklistenerExample”的工程，在 res/layout/activity_main.xml 中添加如实例 3.8 步骤一所示代码。

步骤二： 在 src 目录下打开 MainActivity.java 文件，修改文件，内容如下：

```
public boolean onLongClick(View v)
{
        textView.setText("您选择了"+((Button)v).getText()+",欢迎你！ ");
        return true;
}
```

步骤三：测试接口。

完整程序：

```
public class MainActivity extends ActionBarActivity implements OnLongClickListener {
        Button[] buttons=new Button[3];
            TextView textView;
              public void onCreate(Bundle savedInstanceState){
                super.onCreate(savedInstanceState);
                setContentView(R.layout.activity_main);
              buttons[0]=(Button)this.findViewById(R.id.button1);
              buttons[1]=(Button)this.findViewById(R.id.button2);
              buttons[2]=(Button)this.findViewById(R.id.button3);
              textView=(TextView)this.findViewById(R.id.textView1);
              textView.setTextSize(18);
              for(Button button:buttons){
                      button.setOnLongClickListener(this);}}
              @Override
              public boolean onLongClick(View v){
              textView.setText("您选择了"+((Button)v).getText()+",欢迎你！ ");
              return true;
              }
        }
```

步骤四：运行项目。

三、Button 和 OnTouchListener 触摸事件

对于一个 Android 应用程序来说，事件处理是必不可少的，用户与应用程序之间的交互便是通过事件处理来完成的。当用户与应用程序交互时，一定是通过触发某些事件来完成的，让事件来通知程序应该执行哪些操作，在这个繁杂的过程中主要涉及两个对象，事件源与事件监听器。OnTouchListener 接口是用来处理手机屏幕事件的监听接口，当在 View 的范围内发生触摸按下、抬起或滑动等动作时会触发该事件。该接口中的监听方法声明格式如下：

```
Public Boolean onTouch（View v,MotionEvent event）
```

参数 v：同样为事件源对象

参数 event：为事件封装类的对象，其中封装了触发事件的详细信息，包括事件的类型、触发时间等信息。

【实例 3.10】通过监听接口实现在屏幕上拖动按钮移动，如图 3-9 所示。

步骤一：创建名为“ontouchlistenExample”的工程，在 res/layout/activity_main.xml 中添加如下代码：

```
<AbsoluteLayout xmlns:android="http://schemas.android.com/apk/res/android"
        android:id="@+id/AbsoluteLayout01"
        android:layout_width="fill_parent"
        android:layout_height="fill_parent">
         <Button
          android:id="@+id/button1"
          android:layout_x="100px"
          android:layout_y="100px"
          android:layout_width="wrap_content"
          android:layout_height="wrap_content"
          android:text="Button" />
</AbsoluteLayout>
```

步骤二：在 src 目录下打开 MainActivity.java 文件，修改文件，内容如下：

```
public class MainActivity extends ActionBarActivity {
       final static int WRAP_CONTENT=-2;//表示 WARP_CONTENT 的常量
           final static int X_MODIFY=4; //在非全屏模式下 X 坐标的修正值
           final static int Y_MODIFY=52; //在非全屏模式下 Y 坐标修正值
           int xSpan; //在单击按钮的情况下相对按钮自己坐标系的 X 坐标
           int ySpan; //在单击按钮的情况下相对按钮自己坐标系的 Y 坐标
           protected void onCreate(Bundle savedInstanceState) {
              super.onCreate(savedInstanceState);
              setContentView(R.layout.activity_main);
              Button bok=(Button)this.findViewById(R.id.button1);
              bok.setOnTouchListener(new OnTouchListener(){
              public boolean onTouch(View v,MotionEvent event){
                     switch(event.getAction()){
                     case MotionEvent.ACTION_DOWN://手触按钮按下
                     Button bok1=(Button)findViewById(R.id.button1);
                     bok1.setText("我被按下");
               break;
                     case MotionEvent.ACTION_UP: //手触按钮抬起
                     Button bok2=(Button)findViewById(R.id.button1);
                     //让按钮随着手触一起移动
                     bok2.setText("我已抬起");
                     break;
                     case MotionEvent.ACTION_MOVE://手触按钮移动
                     Button bok=(Button)findViewById(R.id.button1);
                     bok.setText("我已抬起");
                     ViewGroup.LayoutParams lp=new AbsoluteLayout.LayoutParams(WRAP_CONTENT, WRAP_
                     CONTENT,(int)event.getRawX()-xSpan-X_MODIFY,(int)event.getRawY()-ySpan-Y_MODIFY);
                     bok.setLayoutParams(lp);
                     break;              }
                     return true;            } });
       }
}
```

步骤三：运行项目。

3.3.3 图像组件

我们要把一张图片显示在屏幕上，首先需要创建一个图片的对象，在 Android 中，这个对象是 ImageView 对象，ImageView 与 TextView 功能基本类似，主要区别是显示文本资源。在应用开发中图片视图是一个常用且重要的组件。ImageView 是 ImageButton 的父类，可显示一个可以被用户单击的图片按钮，见表 3-9。ImageView 可通过两种方式设置资源：一种是通过 setImageResource 方法来设置要显示的图片资源索引；另一种是使用 setImageBitmap 方法来设置图片资源，详情见表 3-10。

表 3-9　ImageView 控件中常用的属性和方法表

属性	对应方法	解释
android:adjustViewBounds	setAdjustViewBounds(boolean)	设置是否需要 ImageView 调整自己的边界来保证所显示的图片的长宽比例
android:maxHeight	setMaxHeight(int)	ImageView 的最大高度，可选
android:maxWidth	setMaxWidth(int)	ImageView 的最大宽度，可选
android:scaleType	setScaleType(ImageView.ScaleType)	控制图片应如何调整或移动来适合 ImageView 的尺寸
android:src	setImageResource(int)	设置 ImageView 要显示的图片

当然，还可以对图片执行一些其他的操作，比如设置它的 Alpha 值等。这里将直接用一个示例来分析 ImageView 的使用，该示例通过一个线程来不断更新 ImageView 的 Alpha 值，实现效果如图 3-10 所示。当程序运行一段时间后，Alpha 值逐渐变小。

表 3-10　ImageView 控件设置资源的方法

方法	解释
setAlpha(int alpha)	设置 ImageView 的透明度
setImageBitmap(Bitmap bm)	设置 ImageView 所显示的内容为指定 Bitmap 对象
setimageDrawable(Drawable drawable)	设置 ImageView 所显示的内容为指定 drawable
setImageResurce(int resId)	设置 ImageView 所显示的内容为指定 id 的资源
setImageURL(Uri uri)	设置 ImageView 所显示的内容为指定 Uri
setSelected(Boolean selected)	设置 ImageView 的选中状态

【实例 3.11】通过 XML 布局，利用 ImageView 来显示图片，图片下方有四个按钮，分别完成前一张图片、下一张图片、增加图片透明度、降低图片透明度等功能，通过 TextView 来显示文字，布局方式如图 3-10 所示。

图 3-10 ImageView 组件

步骤一：创建名为“ImageViewExample”的工程，在 res/layout/activity_main.xml 中添加如下代码实现界面布局：

```
<LinearLayout xmlns:android="http://schemas.android.com/apk/res/android"
        android:layout_width="fill_parent"
        android:layout_height="fill_parent"
        android:layout_gravity="center_horizontal"
        android:orientation="vertical" >//设置垂直分布的线性布局
          <!—声明 ImageView 控件-->
 <ImageView
        android:id="@+id/iv"
        android:layout_width="wrap_content"
        android:layout_height="wrap_content"
        android:layout_gravity="center_horizontal"
        android:src="@drawable/lzy1" >
 </ImageView>
 <!—声明水平分布的线性布局-->
 <LinearLayout
         android:orientation="horizontal"
         android:layout_width="fill_parent"
         android:layout_height="wrap_content"
         android:layout_gravity="center_horizontal" >
 <!—声明一个“前一张”图片的 Button 控件按钮-->
<Button
       android:id="@+id/previous"
       android:layout_width="wrap_content"
```

```
            android:layout_height="wrap_content"
            android:text="前一张"
            android:layout_gravity="center_horizontal"
            android:layout_weight="1"/>
    <!—声明一个"增加透明度"的 Button 控件按钮-->
    <Button
            android:id="@+id/alpha_plus"
            android:layout_width="wrap_content"
            android:layout_height="wrap_content"
            android:text="增加透明度"
            android:layout_gravity="center_horizontal"
            android:layout_weight="2"/>
    <!—声明一个"降低透明度"的 Button 控件按钮-->
    <Button
            android:id="@+id/alpha_minus"
            android:layout_width="wrap_content"
            android:layout_height="wrap_content"
            android:text="降低透明度"
            android:layout_gravity="center_horizontal"
            android:layout_weight="2"/>
    <!—声明一个"下一张"图片的 Button 控件按钮-->
    <Button
            android:id="@+id/next"
            android:layout_width="wrap_content"
            android:layout_height="wrap_content"
            android:text="下一张"
            android:layout_gravity="center_horizontal"
            android:layout_weight="1"/>
    </LinearLayout>
    </LinearLayout>
```

步骤二：在 src 目录下打开 MainActivity.java 文件，修改文件，内容如下：

```
public class MainActivity extends ActionBarActivity {
    ImageView iv;//ImageView 对象引用
    Button btnNext,btnPrevious,btnAlphaPlus,btnAlphaminus;
    int currImgId=0;//记录当前 ImageView 显示的图片 id
    int alpha=255; //记录当前 ImageView 的透明度
    int [] imgId={R.drawable.lzy1,R.drawable.lzy2,R.drawable.lzy3,R.drawable.lzy4,R.drawable.lzy5};
    //ImageView 显示的图片数组
    private View.OnClickListener myListener=new View.OnClickListener(){
        @Override
        public void onClick(View v){//判断按下的是哪个 Button
            if(v==btnNext){            //下一张图片按钮被按下
                currImgId=(currImgId+1)%imgId.length;
                iv.setImageResource(imgId[currImgId]);
                //设置 ImageView 的显示图片
```

```
                }else if(v==btnPrevious){ //上一张图片按钮被按下
                    currImgId=(currImgId-1+imgId.length)%imgId.length;
                    iv.setImageResource(imgId[currImgId]);
                }//设置 ImageView 的显示图片
                else if(v==btnAlphaPlus){//透明度按钮被按下
                    alpha+=25;
                    if(alpha>255){
                        alpha=0;
                    }
                iv.setAlpha(alpha);//设置 ImageView 透明度
                }
                else
                    if(v==btnAlphaminus)//减少透明度按钮被按下
                        {alpha-=25;
                if(alpha<0)
                    alpha=255;
                iv.setAlpha(alpha);}//设置 ImageView 的透明度
            }
        };

    @Override
    public void onCreate(Bundle savedInstanceState){
        super.onCreate(savedInstanceState);
        setContentView(R.layout.activity_main);
        iv=(ImageView)findViewById(R.id.iv);//获得 ImageView 对象引用
        btnNext=(Button)findViewById(R.id.next); //获得 ImageView 对象引用
        btnPrevious=(Button)findViewById(R.id.previous);
        btnAlphaPlus=(Button)findViewById(R.id.alpha_plus);
        btnAlphaminus=(Button)findViewById(R.id.alpha_minus);
        <!—为 Button 对象设 OnClickListener 监听器-->
        btnNext.setOnClickListener(myListener);
        btnPrevious.setOnClickListener(myListener);
        btnAlphaPlus.setOnClickListener(myListener);
        btnAlphaminus.setOnClickListener(myListener);
        }
    }
```


步骤三：在 res/values/strings.xml 中，修改以下代码。

```
<resources>
    <string name="app_name">imageview</string>
    <string name="hello_world">Hello world!</string>
    <string name="action_settings">Settings</string>
<string name="next">下一张</string>
<string name="previous">上一张</string>
<string name="alpha_plus">透明度增加</string>
<string name="alpha_minus">透明度减小</string>
</resources>
```

步骤四：运行项目。

3.3.4 日期与时间组件

日期和时间是任何手机都会有的基本功能，Android 也不例外。Android 平台使用 DatePicker 来实现日期，用 TimePicker 来实现时间，各类的主要成员方法及说明见表 3-11、表 3-12。Android 平台的日期、时间界面非常漂亮，设置日期和时间的界面如图 3-11 所示。

（a）日期与时间的选择界面

（b）选完时间和日期后的界面

图 3-11　日期与时间组件界面

表 3-11　DatePicker 类主要的成员方法及说明

方法名称	方法说明
getDayOfMonth()	获取日期天数
getMonth()	获取日期月份
getYear()	获取日期年份
init(int year,int monthOfYear,int dayOfMonth,Datepicker.onDateChangedListener onDateChangedListener)	初始化 DatePicker 控件的属性，参数 onDateChangedListener 为监听器对象，负责监听日期数据变化
setEnabled(Boolean enabled)	根据传入的参数设置日期选择控件是否可用
updateDate(int year,int monthOfYear,int dayOfMonth)	根据传入的参数更新日期选择控件的各个属性值

表 3-12　TimePicker 类主要的成员方法及说明

方法名称	方法说明
getCurrentHour()	获取事件选项控件的当前小时，返回 Integer 对象
getCurrentMinute()	获取事件选项控件的当前分钟，返回 Integer 对象
is24HourView()	判断时间选择控件是否为 24 小时制
setCurrentHour(Integer currentHour)	设置时间选择控件的当前小时，传入 Integer 对象
setCurrentMinute(Integer currentMinute)	设置时间选择控件的当前分钟，传入 Integer 对象
setEnabled(boolean enabled)	根据传入的参数设置时间选择控件是否可用
setIs24HourView(Boolean is24HourView)	根据传入的参数设置时间选择控件是否为 24 小时制
setOnTimeChangedListener(TimePicker.OnTimeChangeListener onTimeChangedListener)	为时间选择控件添加 OnTimeChangeListener 监听器

【实例 3.12】设置 Android 的时间和日期界面，如图 3-11 所示。需要在布局文件中定义 DataPicker 和 TimePicker，然后通过 Calendar 类获得系统时间，接着通过 init 方法将日期传递给 DatePicker，并设置 OnDateChangedListener 监听设置时间。

具体实现代码如下所示：

步骤一：创建名为“DateTimeExample”的工程，在 res/layout/activity_main.xml 中添加如下代码：

```
<!--声明一个垂直分布的线性布局-->
    <LinearLayout xmlns:android="http://schemas.android.com/apk/res/android"
        android:layout_width="match_parent"
        android:layout_height="match_parent"
        android:orientation="vertical" >
<!--声明一个 DatePicker 控件-->
<DatePicker
        android:id="@+id/datepicker"
        android:layout_width="wrap_content"
        android:layout_height="wrap_content"
        android:layout_gravity="center_horizontal"/>
        <!--声明一个用于显示日期的 EditText 控件-->
        <EditText android:id="@+id/etdt"
        android:layout_width="fill_parent"
        android:layout_height="wrap_content"
        android:cursorVisible="false"
        android:editable="false"/>
<!--声明一个 TimePicker 控件-->
<TimePicker
        android:id="@+id/timepicker"
        android:layout_width="wrap_content"
```

```
            android:layout_height="wrap_content"
            android:layout_gravity="center_horizontal"/>
    <!--声明一个用于显示时间的 EditText 控件-->
    <EditText android:id="@+id/ettm"
            android:layout_width="fill_parent"
            android:layout_height="wrap_content"
            android:cursorVisible="false"
            android:editable="false"/>
    </LinearLayout>
```

步骤二：在 src 目录下打开 MainActivity.java 文件，修改文件，内容如下：

```
public void onCreate(Bundle savedInstanceState) {
        super.onCreate(savedInstanceState);
        setContentView(R.layout.activity_main);//设置当前屏幕界面
        DatePicker dp=(DatePicker)findViewById(R.id.datepicker);
        TimePicker tp=(TimePicker)findViewById(R.id.timepicker);
        Calendar c=Calendar.getInstance();//获得 Calendar 对象
        int year=c.get(Calendar.YEAR);
        int monthOfYear=c.get(Calendar.MONTH);
        int dayOfMonth=c.get(Calendar.DAY_OF_MONTH);
        dp.init(year, monthOfYear, dayOfMonth, new OnDateChangedListener(){//初始化 DatePicker
          public void onDateChanged(DatePicker view,int year,int monthOfYear,int dayOfMonth){
          flushDate(year,monthOfYear,dayOfMonth);//更新 EditText 所显示内容
          }
        });
        //为 TimePicker 添加监听器
        tp.setOnTimeChangedListener(new OnTimeChangedListener(){
        public void onTimeChanged(TimePicker view,int hourOfDay,int minute){
              flushTime(hourOfDay,minute); /更新 EditText 所显示内容
              }
          });
        }
        //刷新日期 EditText 所显示内容
        public void flushDate(int year,int monthOfYear,int dayOfMonth){
              EditText et=(EditText)findViewById(R.id.etdt);
              et.setText("您选择的日期是："+year+"年"+(monthOfYear+1)+"月"+dayOfMonth+"日。");}
        //刷新时间 EditText 所显示内容
        public void flushTime(int hourOfDay,int minute)
        {
              EditText et=(EditText)findViewById(R.id.ettm);
        et.setText("您选择的时间是："+hourOfDay+"时"+minute+"分。");
        }
}
```

步骤三：运行项目。

3.3.5 菜单组件

菜单可以让应用程序有更完美的用户体验，Android 手机用一个按钮“Menu”专门来显示菜单，所以，当应用程序设置了菜单，我们便可以通过该按键来操作应用程序的菜单选项。Android 在平台下所提供的菜单大体上可以分为三类：选项菜单（Options Menu）、上下文菜单（Context Menu）和子菜单（Submenu）。

一、选项菜单

当 Activity 在前台运行时，如果用户按下手机上的 Menu 键，此时就会在屏幕底部弹出相应的选择菜单。但这个功能是需要开发人员编程来实现的，如果在开发应用程序时没有实现该功能，那么程序运用时按下手机上的 Menu 键是不会起作用的。

对于携带图标的选项菜单，每次最多只能显示 6 个菜单，但多于 6 个时，将只显示前 5 个和一个扩展菜单选项，单击扩展菜单选项将会弹出其余的菜单项。扩展菜单项中不显示图标，可以显示单选按钮及复选框，具体见表 3-13。

表 3-13 选项菜单相关的回调方法及说明

方法名	描述
onCreateoptionsMenu(Menu menu)	初始化选项菜单，该方法只在第一次显示菜单时调用，如果每次显示菜单时更新菜单项，则需要重写 OnPrepareOptionsMenu(Menu)
public Boolean onOptionsItemselected(Menuitem item)	当选项菜单中某个选项被选中时调用该方法，默认的是一个返回 false 的空实现
public void onOptionsMenuclosed(Menu menu)	当选项菜单关闭时（或者由于用户按下了返回键或者选择了某个菜单选项）调用该方法。
public boolean onPrepareOptionsMenu(Menu menu)	为程序准备选项菜单，每次选项菜单显示前会调用该方法，可以通过该方法设置某些菜单项可用或不可用或者修改菜单项的内容。重写该方法时需要返回 true，否则选项菜单将不会显示。

二、上下文菜单

上下文菜单（ContextMenu）的使用，ContextMenu 继承自 Menu。上下文菜单不同于选项菜单，选项菜单服务于 Activity，而上下文菜单是注册到某个 View 对象上的，具体参数及说明见表 3-14。如果一个 View 对象注册了上下文菜单，用户可以通过长按（约 2s）该 View 对象显示上下文菜单。

上下文菜单不支持快捷键（shortcut），其菜单选项也不能附带图标，但是可以为上下文菜单的标题指定图标。

表 3-14　Activity 类中与 ContextMenu 相关的方法及说明

方法名称	参数说明	方法说明
onCreateContextMenu(ContextMenu menu,View v,ContextMenu.ContextMenuInfo menuInfo)	Menu:创建的上下文菜单 V:上下文菜单依附的 View 对象；MenuInfo：上下文菜单需要额外显示的信息	每次为 View 对象呼出上下文菜单时都将调用该方法
onContextItemSelected(MenuItem item)	Item:被选中的上下文菜单选项	当用户选择了上下文菜单选项后调用该方法进行处理
onContextMenuaClose(Menu menu)	menu：被关闭的上下文菜单	当上下文菜单被关闭时调用该方法
registerForContextMenu(View view)	View：要显示上下文菜单的 View 对象	为指定的 View 对象注册一个上下文菜单

（a）菜单界面

（b）长按菜单后的菜单界面

图 3-12　上下文菜单

【实例 3.13】通过实例设计菜单如图 3-12 所示，具体学习上下文菜单的使用方法。

步骤一：创建名为“TextViewExample”的工程，在 res/layout/activity_main.xml 中添加如下代码：

```
<?xml version="1.0" encoding="utf-8"?>
<LinearLayout xmlns:android="http://schemas.android.com/apk/res/android"
  android:orientation="vertical"
```

```
        android:layout_width="fill_parent"
        android:layout_height="fill_parent">
    <TextView
        android:layout_width="fill_parent"
        android:layout_height="wrap_content"
        android:text="长时间按住，显示菜单哦！">
    </TextView>
    <TextView
        android:id="@+id/textview1"
        android:layout_width="fill_parent"
        android:layout_height="wrap_content"
        android:text="上下文菜单">
    </TextView>
    </LinearLayout>
```

步骤二： 在 src 目录下打开 MainActivity.java 文件，修改文件，内容如下：

```
public class MainActivity extends ActionBarActivity {
    private TextView textview1;
    private ContextMenuInfo contextmenuInfo;
    @Override
    public void onCreate(Bundle savedInstanceState)
    {
        super.onCreate(savedInstanceState);
        try{
            setContentView(R.layout.activity_main);
            init();
            createComponent();
            register();
        }
        catch(Exception e){
            Toast.makeText(this,"异常错误"+e.toString(),Toast.LENGTH_LONG).show();
        }
        finally{}
    }
    public void init(){
        textview1=null;
    }
    public void createComponent(){
        textview1=(TextView)this.findViewById(R.id.textview1);
    }
    public void register(){
        this.registerForContextMenu(textview1);
    }
    @Override
    public void onCreateContextMenu(ContextMenu contextMenu,View view,ContextMenuInfo contextMenuInfo)
    {
```

```
            super.onCreateContextMenu(contextMenu, view, contextmenuInfo);
            if(view==textview1)
            {
                contextMenu.setHeaderIcon(R.drawable.icon);
                contextMenu.setHeaderTitle("My menu");
                contextMenu.add(1,0,0,"菜单 1");
                contextMenu.add(1,1,1,"菜单 2");
                contextMenu.add(1,1,2,"菜单 3");
                contextMenu.add(1,1,2,"菜单 4");
                contextMenu.add(1,1,2,"菜单 5");
                contextMenu.add(1,1,2,"菜单 6");
            }
        }
    }
```

步骤三：运行项目。

三、子菜单（SubMenu）

子菜单继承自 Menu，每个 SubMenu 实例代表一个子菜单，SubMenu 中常用的方法及说明见表 3-15。

表 3-15　Submenu 中常用方法及说明

方法名称	参数说明	方法说明
setHeaderIcon(Drawable icon)	icon:标题图标 Drawable 对象	设置子菜单的标题图标
setHeaderIcon(int iconRes)	iconRes:标题图标的资源 id	
setHeaderTitle(int titleRes)	titleRes：标题文本的资源 id	设置子菜单的标题
setHeaderIcon(charSequence title)	title：标题文本对象	
setIcon(Drawable icon)	icon：图标 Drawable 对象	设置子菜单在父菜单中显示的标题
setIcon(int iconRes)	iconRes：图标资源 id	
SetHeaderView(View view)	view：用于子菜单标题的 view 对象	设置指定 View 对象作为子菜单图标

【实例 3.14】设置简单的子菜单，如图 3-13 所示。

方法：首先需要通过 onCreateOptionsMenu 创建菜单，然后需要对其能够触发的事件进行监听，这样才能在事件监听 onOptionsItemSelected 中根据不同的菜单选项来执行不同的任务。当然，可以通过 XML 布局来实现，也可以通过 menu.add 方法来实现，下面分两种方式来实现菜单效果，菜单效果如图 3-14 所示。

（a）长按“子菜单”

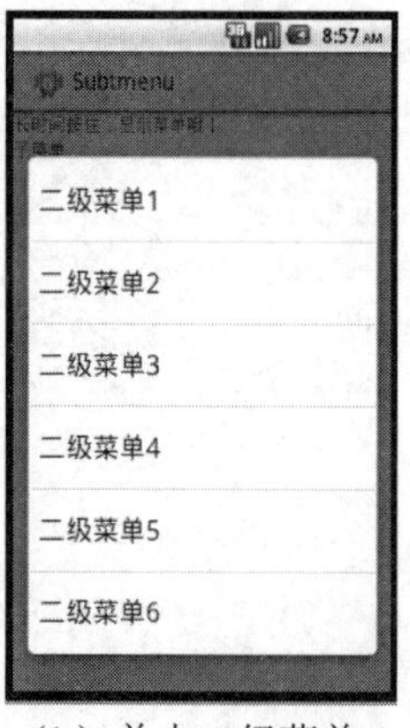

（b）单击二级菜单 1

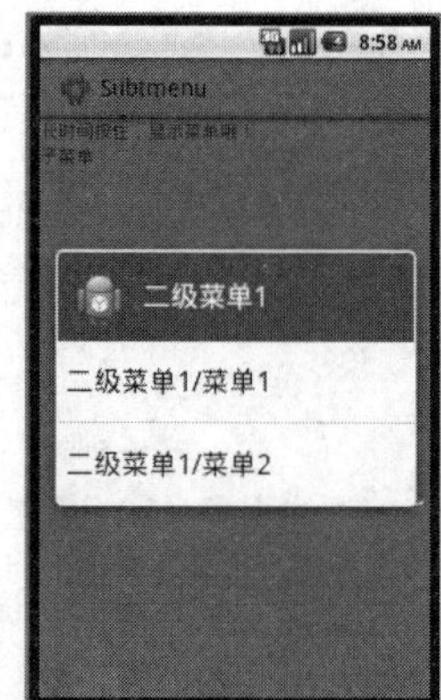

（c）显示子菜单

图 3-13　子菜单的功能实现

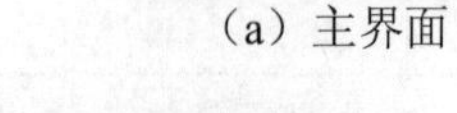

（a）主界面

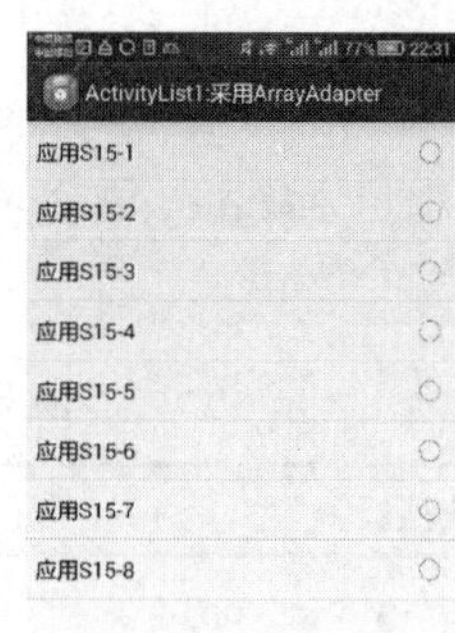

（b）ActivityList1 运行界面

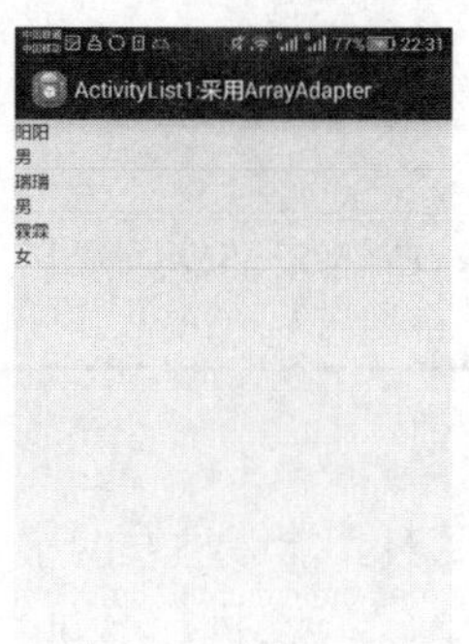

（c）ActivityList2 运行界面

（d）ActivityList3 运行界面

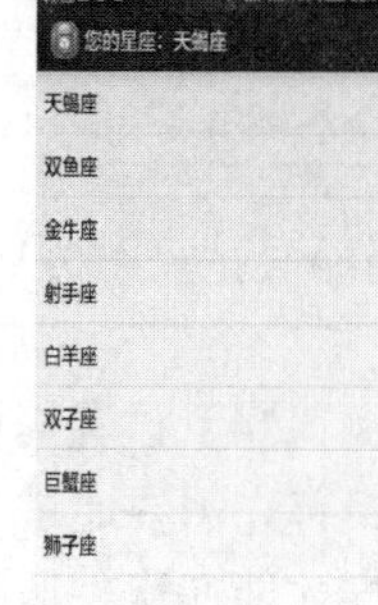

（e）ActivityList4 运行界面

图 3-14　ListView 测试界面

步骤一：创建名为“SubmenuExample”的工程，在 res/layout/activity_main.xml 中添加如下代码：

```
<?xml version="1.0" encoding="utf-8"?>
<LinearLayout
xmlns:android="http://schemas.android.com/apk/res/android"
 android:orientation="vertical"
     android:layout_width="fill_parent"
     android:layout_height="fill_parent">
        <TextView
         android:layout_width="fill_parent"
         android:layout_height="wrap_content"
         android:text="长时间按住，显示菜单哦！">
    </TextView>
        <TextView
        android:id="@+id/textview2"
        android:layout_width="fill_parent"
        android:layout_height="wrap_content"
        android:text="子菜单">
        </TextView>
   </LinearLayout>
```

步骤二：在 src 目录下打开 MainActivity.java 文件，修改文件，内容如下：

```
Public class mainactivity extends actionbaractivity {
Private textview textview2;
      Private contextmenuinfo contextmenuinfo;
       @Override
       Public void oncreate(Bundle savedinstancestate)
        {
          Super.oncreate(savedinstancestate);
          Try{
               Setcontentview(R.layout.activity_main);
               Init();
               Createcomponent();
               Register();
          }
          Catch(Exception e){
          Toast.maketext(this,"异常错误"+e.tostring(),Toast.LENGTH_LONG).show();
          }
          Finally{}
     }
     Public void init(){
          Textview2=null;
     }
     Public void createcomponent(){
     Textview2=(textview)this.findviewbyid(R.id.textview2);
     }
     Public void register(){
```

```
            This.registerforcontextmenu(textview2);
        }
        @Override
        Public void oncreatecontextmenu(contextmenu contextmenu,View view,contextmenuinfo contextmenuinfo)
        {
        Super.oncreatecontextmenu(contextmenu, view, contextmenuinfo);
        If(view==textview2)
        {submenu submenu1=contextmenu.addsubmenu("二级菜单 1");
         Submenu1.setheadericon(R.drawable.icon);
         Submenu1.add(0,0,0,"二级菜单 1/菜单 1");
         Submenu1.add(0,1,1,"二级菜单 1/菜单 2");
         Submenu1.setgroupcheckable(1,true,true);
         Submenu submenu2=contextmenu.addsubmenu("二级菜单 2");
         Submenu2.setheadericon(R.drawable.icon);
         Submenu2.add(1,0,0,"二级菜单 2/菜单 1");
         Submenu2.add(1,1,1,"二级菜单 2/菜单 2");
         Submenu2.setgroupcheckable(1,true,true);
         Submenu submenu3=contextmenu.addsubmenu("二级菜单 3");
         Submenu3.setheadericon(R.drawable.icon);
         Submenu3.add(1,0,0,"二级菜单 3/菜单 1");
         Submenu3.add(1,1,1,"二级菜单 3/菜单 2");
         Submenu3.setgroupcheckable(1,true,true);
         Submenu submenu4=contextmenu.addsubmenu("二级菜单 4");
         Submenu4.setheadericon(R.drawable.icon);
         Submenu4.add(1,0,0,"二级菜单 4/菜单 1");
         Submenu4.add(1,1,1,"二级菜单 4/菜单 2");
         Submenu4.setgroupcheckable(1,true,true);
         Submenu submenu5=contextmenu.addsubmenu("二级菜单 5");
         Submenu5.setheadericon(R.drawable.icon);
         Submenu5.add(1,0,0,"二级菜单 5/菜单 1");
         Submenu5.add(1,1,1,"二级菜单 5/菜单 2");
         Submenu5.setgroupcheckable(1,true,true);
         Submenu submenu6=contextmenu.addsubmenu("二级菜单 6");
         Submenu6.setheadericon(R.drawable.icon);
         Submenu6.add(1,0,0,"二级菜单 6/菜单 1");
         Submenu6.add(1,1,1,"二级菜单 6/菜单 2");
         Submenu6.setgroupcheckable(1,true,true);
        }
    }
}
```

步骤三：运行项目。

3.3.6 列表组件和相关事件

在 Android 中，ListView 用来显示一个列表的控件。ListView 显示的数据并没有直接存储到 ListView 中，而是通过一个 Adapter 对象封装。使用下列方法为 ListView 指定 Adapter 对象：

```
public void setAdapter (ListAdapter adapter)
```

ListAdapter 有很多子类，其中 ArrayAdapter 封装数组，是较常用的一种 Adapter，构造方法如下：

```
ArrayAdapter(Context context,int resource,T[]objects)
```

Context 是 Adapter 所在上下文，往往使用 Activity 对象；resource 指用来显示数据项的布局文件，可以使用自定义的文件，也可以使用系统提供的文件；objects 是该 Adapter 封装的数组。

ListActivity 是 Activity 类的子类，能够更为便捷使用 ListView。然而，并不是只有 ListActivity 中可以使用 ListView，任何 Activity 都可以使用 ListView。注意：ListActivity 对象持有一个 ListView 对象，该 ListView 的 ID 必须是@android：id/list。

ListView 可以将 Adapter 对象封装的数据源以列表方式显示，在实际应用中，往往需要对单击数据项的事件进行监听，与 ListView 列表项有关的事件有两种，当鼠标滚动时会触发 setOnItemselectedListener 事件，单击时则会产生 setOnItemClickListener 事件。本节主要涉及 setOnItemClickListener 事件，设置列表组件的界面如图 3-14 所示。

【实例 3.15】设置列表实例如图 3-14 所示。

步骤一：创建名为“listViewExample”的工程，在 res/layout/activity_main.xml 中添加如下代码：

```
<LinearLayout xmlns:android="http://schemas.android.com/apk/res/android"
        android:layout_width="match_parent"
        android:layout_height="match_parent"
        android:orientation="vertical" >
        <Button
            android:id="@+id/list_view_button_1"
            android:layout_width="wrap_content"
            android:layout_height="wrap_content"
            android:text="列表" />
        <Button
            android:id="@+id/list_view_button_2"
            android:layout_width="wrap_content"
            android:layout_height="wrap_content"
            android:text="列表的显示" />
        <Button
            android:id="@+id/list_view_button_3"
            android:layout_width="wrap_content"
            android:layout_height="wrap_content"
            android:text="没有数据的列表" />
        <Button
            android:id="@+id/list_view_button_4"
            android:layout_width="wrap_content"
            android:layout_height="wrap_content"
            android:text="对列表进行单击操作" />
    </LinearLayout>
```

步骤二：在 layout 文件夹中新建一个 xml 文件，命名为 list_item.xml，代码如下：

```
<?xml version="1.0" encoding="utf-8"?>
<LinearLayout xmlns:android="http://schemas.android.com/apk/res/android"
      android:layout_width="match_parent"
      android:layout_height="match_parent"
      android:orientation="vertical" >
      <TextView
            android:id="@+id/mview1"
            android:layout_width="wrap_content"
            android:layout_height="wrap_content"
            android:text="" />
      <TextView
            android:id="@+id/mview2"
            android:layout_width="wrap_content"
            android:layout_height="wrap_content"
            android:text="" />
</LinearLayout>
```

步骤三：在 layout 文件夹中新建一个 xml 文件，命名为 list3.xml，代码如下：

```
<?xml version="1.0" encoding="utf-8"?>
<LinearLayout xmlns:android="http://schemas.android.com/apk/res/android"
      android:layout_width="match_parent"
      android:layout_height="match_parent"
      android:orientation="vertical" >
      <TextView
            android:id="@+id/tv1"
            android:layout_width="wrap_content"
            android:layout_height="wrap_content"
            android:text="对不起，没有数据显示" />
      <ListView
            android:id="@+id/android:list"
            android:layout_width="286dp"
            android:layout_height="182dp" >
      </ListView>
</LinearLayout>
```

请注意：listView 中 android:id="@+id/android:list"的修改。

步骤四：在 src 目录下打开 MainActivity.java 文件，修改文件，内容如下：

```
public class MainActivity extends ActionBarActivity {
        OnClickListener listener1=null;
        OnClickListener listener2=null;
        OnClickListener listener3=null;
        OnClickListener listener4=null;
        Button button1;
        Button button2;
        Button button3;
        Button button4;
        @Override
```

```
        protected void onCreate(Bundle savedInstanceState) {
                super.onCreate(savedInstanceState);
                prepareListeners();
                setContentView(R.layout.activity_main);
                button1=(Button)findViewById(R.id.list_view_button_1);
                button1.setOnClickListener(listener1);
                button2=(Button)findViewById(R.id.list_view_button_2);
                button2.setOnClickListener(listener2);
                button3=(Button)findViewById(R.id.list_view_button_3);
                button3.setOnClickListener(listener3);
                button4=(Button)findViewById(R.id.list_view_button_4);
                button4.setOnClickListener(listener4);
                }
        private void prepareListeners(){
                listener1=new OnClickListener(){
        public void onClick(View v)
                {Intent intent1=new Intent(MainActivity.this,ActivityList1.class);
                startActivity(intent1);}};
                listener2=new OnClickListener(){
        public void onClick(View v){
                Intent intent2=new Intent(MainActivity.this,ActivityList2.class);
                startActivity(intent2);}};
        listener3=new OnClickListener(){
        public void onClick(View v){
                Intent intent3=new Intent(MainActivity.this,ActivityList3.class);
                startActivity(intent3);}};
        listener4=new OnClickListener(){
        public void onClick(View v){
                Intent intent4=new Intent(MainActivity.this,ActivityList4.class);
                startActivity(intent4);}};
                }
        }
```

步骤五：在 src 目录下新建四个 java 文件，分别命名为 ActivityList1.java、ActivityList2.java、ActivityList3.java、ActivityList4.java，代码分别如下：

```
ActivityList1.java：
public class ActivityList1 extends ActionBarActivity {
ListView listView;
private String[] data={"应用 S15-1","应用 S15-2","应用 S15-3","应用 S15-4","应用 S15-5","应用 S15-6",
"应用 S15-7","应用 S15-8"};
@Override
public void onCreate(Bundle savedInstanceState){
        super.onCreate(savedInstanceState);
        listView=new ListView(this);
listView.setAdapter(new ArrayAdapter<String>(this,android.R.layout.simple_list_item_single_choice,data));
listView.setItemsCanFocus(true);
```

```
listView.setChoiceMode(ListView.CHOICE_MODE_MULTIPLE);
setContentView(listView);
}
}
ActivityList2.java：
public class ActivityList2 extends ActionBarActivity{
        private List<Map<String,Object>>data;
        private ListView listView=null;
        @Override
        public void onCreate(Bundle savedInstanceState){
                super.onCreate(savedInstanceState);
                PrepareData();
                listView=new ListView(this);
                SimpleAdapter adapter=new SimpleAdapter(this,data,R.layout.list_item,new String[]
                {"姓名","性别"},new int[]{R.id.mview1,R.id.mview2});
                listView.setAdapter(adapter);
                setContentView(listView);
                OnItemClickListener listener=new OnItemClickListener(){
        public void onItemClick(AdapterView<?>parent,View view,int position,long id){
                        setTitle(parent.getItemAtPosition(position).toString());
                        }
                };
                listView.setOnItemClickListener(listener);
        }
        private void PrepareData(){
                data=new ArrayList<Map<String,Object>>();
                Map<String,Object>item;
                item=new HashMap<String,Object>();
                item.put("姓名", "阳阳");
                item.put("性别", "男");
                data.add(item);
                item=new HashMap<String,Object>();
                item.put("姓名", "瑞瑞 ");
                item.put("性别", "男");
                data.add(item);
                item=new HashMap<String,Object>();
                item.put("姓名", "霖霖");
                item.put("性别", "女");
                data.add(item);
        }
}
ActivityList3.java：
public class ActivityList3 extends ListActivity{
        private String[] data={"l","p"};
        public void onCreate(Bundle savedInstanceState){
                super.onCreate(savedInstanceState);
```

```
            setContentView(R.layout.list3);
            setListAdapter(new ArrayAdapter<String>(this,android.R.layout.simple_list_item_1,data));
            //其中 simple_list_item_1 是系统自带的布局文件
        }
        protected void onListItemClick(ListView listView,View v,int position,long id){
            super.onListItemClick(listView,v,position,id);
            setTitle(listView.getItemAtPosition(position).toString());
        }
}
ActivityList4.java：
public class ActivityList4 extends Activity{
        ListView listView;
        private String[] data={"天蝎座","双鱼座","金牛座","射手座","白羊座","双子座","巨蟹座","狮子座"};
        @Override
        public void onCreate(Bundle savedInstanceState){
            super.onCreate(savedInstanceState);
            listView=new ListView(this);
        listView.setAdapter(new ArrayAdapter<String>(this,android.R.layout.simple_list_item_1,data));
        setContentView(listView);
        OnItemClickListener itemClickListener=new OnItemClickListener(){
        @Override
        public void onItemClick(AdapterView<?>parent,View arg1,int position,long arg3){
            setTitle("您的星座："+parent.getItemAtPosition(position).toString());
            }
        };
        listView.setOnItemClickListener(itemClickListener);
        }
}
```

步骤六：修改 AndroidManifest.xml 文件，修改代码如下：

```
<?xml version="1.0" encoding="utf-8"?>
<manifest xmlns:android="http://schemas.android.com/apk/res/android"
    package="com.example.listview2"
    android:versionCode="1"
    android:versionName="1.0" >
    <uses-sdk
        android:minSdkVersion="10"
        android:targetSdkVersion="19" />
    <application
        android:allowBackup="true"
        android:icon="@drawable/ic_launcher"
        android:label="@string/app_name"
        android:theme="@style/AppTheme" >
     <activity
         android:name="com.example.listview2.MainActivity"
         android:label="@string/app_name" >
```

```
            <intent-filter>
            <action android:name="android.intent.action.MAIN" />
      <category android:name="android.intent.category.LAUNCHER" />
            </intent-filter>
        </activity>
         <activity android:name=".ActivityList1"
          android:label="ActivityList1:采用 ArrayAdapter">
          </activity>
            <activity android:name=".ActivityList2"
              android:label="ActivityList1:采用 ArrayAdapter">
          </activity>
     <activity android:name=".ActivityList3"
            android:label="演示 ListActivity">
   </activity>
     <activity android:name=".ActivityList4"
      android:label="如何处理对列表的单击操作">
 </activity>
       </application>
 </manifest>
```

步骤七：运行项目。

3.3.7 对话框组件

在 Android 中，实现对话框可以使用 AlterDialog.Builder，还可以用自定义对话框。如果对话框设置了按钮，那么需要对其设置事件监听 OnClickListener。设置对话框组件的界面如图 3-15 所示。

（a）“登录提示”界面

（b）登录界面

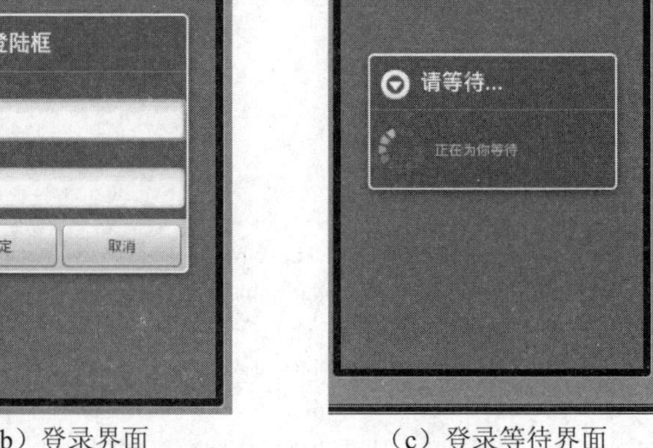

（c）登录等待界面

图 3-15 登录界面

使用 AlterDialog 创建对话框常用方法及说明见表 3-16。

表 3-16 AlterDialog 创建对话框常用方法及说明

方法名称	参数说明
setTitle()	给对话框设置 title
setIcon()	给对话框设置图标
setMessage()	设置对话框的提示信息
setItems()	设置对话框要显示一个 list，一般用于显示几个命令
setSingleChoiceItems()	设置对话框显示一个单选 List
setMultiChoiceItems()	用来设置对话框显示一系列的复选框
setPositiveButton()	给对话框添加“Yes”按钮
setNegativeButton()	给对话框添加“No”按钮

【实例 3.16】设置登录界面，要求单击“确定”，可以转移到我们设计的对话框界面，单击“退出”可以退出应用程序界面。

实现对话框的思路是：首先使用 AlertDialog。Builder()生成默认对话框，当单击确定时，会调用 setPositiveButton 中的 DialogInterface.OnClicklistener()方法，使用 layoutInflater 获取登录框自定义的布局文件，然后新建一个 AlertDialog 对话框，并使用上面自定义的布局文件，如图 3-15 所示。当输入用户名、密码登录后，创建一个 ProgressDialog，建立一个线程，3s 后消失。

步骤一：创建名为“DialogExample”的工程，在 res/layout/中新建 dialog.xml，并添加如下代码：

```
<?xml version="1.0" encoding="utf-8"?>
<LinearLayout xmlns:android="http://schemas.android.com/apk/res/android"
        android:layout_width="fill_parent"
        android:layout_height="fill_parent"
        android:orientation="vertical" >
         <TextView
          android:layout_width="fill_parent"
          android:layout_height="wrap_content"
          android:id="@+id/password"
          />
        <EditText
          android:id="@+id/username"
          android:layout_width="fill_parent"
          android:layout_height="wrap_content"
          android:singleLine="true"
          android:text=""/>
```

```
            <TextView
                android:layout_width="fill_parent"
                android:layout_height="wrap_content"
                android:id="@+id/password"/>
            <EditText
                android:id="@+id/password"
                android:layout_width="fill_parent"
                android:layout_height="wrap_content"
                android:password="true"
                android:singleLine="true"
                android:text=""/>
    </LinearLayout>
```

步骤二：在 src 目录下打开 MainActivity.java 文件，修改文件，内容如下：

```
public class MainActivity extends ActionBarActivity {
        ProgressDialog m_Dialog;
        @Override
        public void onCreate(Bundle savedInstanceState)
        {
                super.onCreate(savedInstanceState);
                setContentView(R.layout.activity_main);
                Dialog dialog=new AlertDialog.Builder(MainActivity.this)
                .setTitle("登录提示")     //设置标题
                .setMessage("这里需要登录！")     //设置内容
                .setPositiveButton("确定",new DialogInterface.OnClickListener() {
                //设置确定按钮

                public void onClick(DialogInterface dialog, int whichButton)
                 {<!--单击“确定”转向登录界面-->
                LayoutInflater factory=LayoutInflater.from(MainActivity.this);
                <!--得到自定义对话框-->
                final View DialogView=factory.inflate(R.layout.dialog,null);
                <!--创建对话框-->
                AlertDialog dlg=new AlertDialog.Builder(MainActivity.this)
                .setTitle("登录框")
                .setView(DialogView) //设置自定义对话框
                .setPositiveButton("确定",new DialogInterface.OnClickListener()
                //设置“确定按钮”及事件监听
                {public void onClick(DialogInterface dialog,int whichButton)
                {//输入完成后单击“确定”开始登录
                        m_Dialog=ProgressDialog.show(MainActivity.this,"请等待...","正在为你等待",true);
                new Thread(){
                public void run(){
                                try{sleep(3000);}
```

```
                        catch(Exception e)
                        {
                            e.printStackTrace();}
                            finally
                            {
                            m_Dialog.dismiss();//登录结束，取消 m_Dialog 对话框}
                            }
                        }.start();
                            }
                        })
                        .setNegativeButton("取消",new DialogInterface.OnClickListener()
                        {public void onClick(DialogInterface dialog,int whichbutton)
                        {//单击“取消”按钮退出程序
                            MainActivity.this.finish();
                        }}
                                )
                                .create();//创建
                                dlg.show();//显示
            }
            }).setNeutralButton("退出",new DialogInterface.OnClickListener(){
            public void onClick(DialogInterface dialog,int whichButton)
            {MainActivity.this.finish();//单击“退出”按钮，退出程序

            }
    }).create();//创建按钮
            dialog.show();//显示对话框
    }}
```

步骤三：运行项目。

3.3.8 进度条组件

当一个应用程序在后台执行时，前台界面就不会有什么内容，这时用户根本不知道程序是否在执行、执行进度如何、应用程序是否遇到错误终止等，这时需要使用进度条来提示用户后台程序执行的进度。Android 系统提供了两大类进度条样式，长条形进度条（progressBarStyleHorizontal）和圆形进度条（progressBarStyleLarge）。进度条用处很多，比如，应用程序加载资源和网络连接时，可以提示用户稍等，这一类进度条只是代表应用程序某一部分程序的执行情况，而整个应用程序执行情况，则可以通过应用程序标题栏来显示一个进度条，这就需要先对窗口的显示风格进行设置“requestWindowFeature(Widow.FEATURE_PROGRESS);”。进度条组件的界面如图 3-16 所示。

（a）单击“开始”按钮

（b）长形和圆形进度条状态

图 3-16　进度条

【实例 3.17】设置长形和圆形、标题栏进度条的使用。

步骤一：创建名为“ProgressBarExample”的工程，在 res/layout/activity_main.xml 中添加如下代码：

```
<?xml version="1.0" encoding="utf-8"?>
        <LinearLayout    xmlns:android="http://schemas.android.com/apk/res/android"
          android:layout_width="match_parent"
          android:layout_height="match_parent"
          android:orientation="vertical" >
          <TextView
                android:layout_width="fill_parent"
                android:layout_height="wrap_content"
                android:text="@string/hello_world" />
          <ProgressBar
                android:id="@+id/ProgressBar01"
                style="?android:attr/progressBarStyleHorizontal"
//设置风格为长形的进度条
                android:layout_width="200dp"
                android:layout_height="wrap_content"
                android:visibility="gone"/>//设置长形进度条
         <ProgressBar android:id="@+id/progressBar2"
                style="?android:attr/progressBarStyleLarge"
                android:layout_width="wrap_content"
                android:layout_height="wrap_content"
                android:max="100"              //最大进度值为 100
                android:progress="50"     //初始化的进度值
                android:secondaryProgress="70"
```

```
//初始化的底层第二个进度值
        android:visibility="gone"/>//设置圆形进度条
    <Button
        android:id="@+id/button1"
        android:layout_width="wrap_content"
        android:layout_height="wrap_content"
        android:text="开始" />
</LinearLayout>
```

步骤二：在 src 目录下打开 MainActivity.java 文件，修改文件，内容如下：

```
public class MainActivity extends ActionBarActivity {
    private ProgressBar m_ProgressBar;//声明 ProgressBar 对象
    private ProgressBar m_ProgressBar2;
    private Button mButton01;
    protected static final int GUI_STOP_NOTIFIER=0x108;
    protected static final int GUI_THREADING_NOTIFIER=0x109;
    public int intCounter=0;
    @Override
    public void onCreate(Bundle savedInstanceState)
    {
        super.onCreate(savedInstanceState);
        //设置窗口模式，显示进度条在标题栏
        requestWindowFeature(Window.FEATURE_PROGRESS);
        setProgressBarVisibility(true);
        setContentView(R.layout.activity_main);
        m_ProgressBar=(ProgressBar) findViewById(R.id.ProgressBar01);
        m_ProgressBar2=(ProgressBar) findViewById(R.id.progressBar2);
        mButton01=(Button)findViewById(R.id.button1);
        m_ProgressBar.setIndeterminate(false);
        m_ProgressBar2.setIndeterminate(false);
        <!—当按下按钮时开始执行进度条-->
        mButton01.setOnClickListener(new Button.OnClickListener()
        {@Override
    public void onClick(View v){
        m_ProgressBar.setVisibility(View.VISIBLE);
        m_ProgressBar2.setVisibility(View.VISIBLE);
        m_ProgressBar.setMax(100);//设置 ProgressBar 的最大值
        m_ProgressBar.setProgress(0); //设置 ProgressBar 当前值
        m_ProgressBar2.setProgress(0);
        <!—通过线程改变 ProgressBar 的值-->
            new Thread(new Runnable(){
                public void run(){
                    for(int i=0;i<10;i++){
                        try
                        {intCounter=(i+1)*20;
                         Thread.sleep(1000);
                         if(i==4){
                            Message m=new Message();
```

```
                        m.what=MainActivity.GUI_STOP_NOTIFIER;
                        MainActivity.this.myMessageHandler.sendMessage(m);
                    break;}
                    else
                    {
                        Message m=new Message();
                        m.what=MainActivity.GUI_THREADING_NOTIFIER;
                        MainActivity.this.myMessageHandler.sendMessage(m)}
                    }
                    catch(Exception e){
                    e.printStackTrace();
                    }
                }
            }
        }).start();
    }
        }
    );
}
Handler myMessageHandler=new Handler(){
    public void handleMessage(Message msg)
    {
        switch(msg.what)
        {//ProgressBar 已经是最大值
        case MainActivity.GUI_STOP_NOTIFIER:
        m_ProgressBar.setVisibility(View.GONE);
        m_ProgressBar2.setVisibility(View.GONE);
        Thread.currentThread().interrupt();
        break;
        case MainActivity.GUI_THREADING_NOTIFIER:
        if(!Thread.currentThread().isInterrupted())
        {   // 改变 ProgressBar 的当前值
            m_ProgressBar.setProgress(intCounter);
            m_ProgressBar2.setProgress(intCounter);
            //设置标题栏中前景的一个进度条进度值
            setProgress(intCounter*100);
            //设置标题栏中后面的一个进度条进度值
            setSecondaryProgress(intCounter*100);
        }
        break;
        }
        super.handleMessage(msg);
    }};
}
```

步骤三：运行项目。

3.4 界面资源的定义与使用

巧妇难为无米之炊，无论做什么事都需要资源作为支持，Android 也不例外。Android 系统把字符串、颜色、图片等都看作资源来使用。

在 Android 项目初创完成之后，我们可以在项目中找到 assets 和 res 文件夹，这两个文件夹用来存放资源，assets 文件夹中的资源是通过二进制流形式读取的原生文件，不能直接读取应用，故使用较少。res 文件夹中的资源是我们经常要用到的资源，写在“.xml”文件当中，并且在 R.java 文件中也都为其生成 ID 以供使用。

除此之外，系统本身也会提供一些资源供用户使用，下面来说一下这些资源。

3.4.1 系统资源

系统资源位于 SDK 内部的文件夹中，具体位置为 platforms\android 版本\data\res，这些资源可直接使用，但在使用系统资源时需要通过“@android:”进行引用。

【实例 3.18】如图 3-17 所示，为按钮添加一个背景图片。

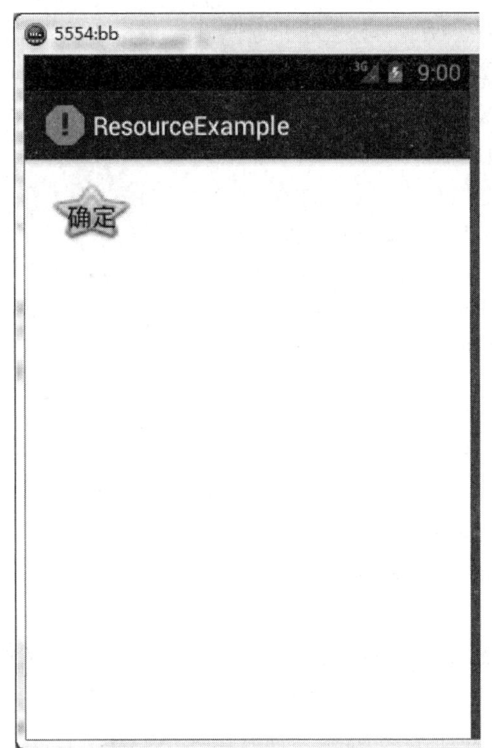

图 3-17　按钮添加背景图片

创建名为“ResourceExample”的工程，在 res/layout/activity_main.xml 中添加如下代码：

```
<LinearLayout xmlns:android="http://schemas.android.com/apk/res/android"
      xmlns:tools="http://schemas.android.com/tools"
      android:layout_width="match_parent"
      android:layout_height="match_parent"
      android:orientation="vertical"
      tools:context=".MainActivity" >
<Button
      android:layout_width="wrap_content"
      android:layout_height="wrap_content"
      android:text="确定"
      android:background="@android:drawable/btn_star_big"/>
</LinearLayout>
```

在 Button 控件中 background 表示背景，其值@android:drawable/btn_star_big 便是我们要添加的背景图片，其中@android:为我们使用系统资源时的固定格式，drawable 为资源类型，btn_star_big 为资源 ID。

3.4.2 字符串资源（String）

字符串资源指的就是字符串，只是为了方便修改字符串内容，android 将所用到的字符串放在了一起。字符串资源位于项目中，具体的位置为 res\values\strings.xml，字符串资源会写在文件 strings.xml 中，打开文件便可看到。

string.xml 文件中的内容

```
<resources>
<string name="app_name">Android 资源</string>
<string name="hello_world">Hello world!</string>
<string name="menu_settings">Settings</string>
<string name="title_activity_main">Android 资源</string>
<string name="string1">你好欢迎来到应用系！</string>
<string name="string2">你好欢迎再次来到应用系！</string>
</resources>
```

在这个文件中所有的字符串要位于<resources></resources>之中，对于字符串内容我们以字符串 string1 为例，<string name="string1">中 name 为字符串的名称属性，string1 为这个字符串名称属性的值，在<string></string >之中的部分即“你好欢迎来到应用系！”为字符串的实际内容。

对于字符串的引用我们要分为两种情况，第一种是在 xml 文件中进行引用，第二种是在 java 文件中进行引用。首先我们来看在 xml 文件中进行引用。

【实例 3.19】如图 3-18 所示，显示字符串资源“你好欢迎来到应用系！”。

图 3-18　显示字符串

activity_main.xml 文件中的内容，即 activity 的布局文件内容。

```
<LinearLayout xmlns:android="http://schemas.android.com/apk/res/android"
    xmlns:tools="http://schemas.android.com/tools"
    android:layout_width="match_parent"
    android:layout_height="match_parent"
    android:orientation="vertical" >
<TextView
        android:id="@+id/lblTitle1"
        android:layout_width="fill_parent"
        android:layout_height="wrap_content"
        android:text="@string/string1"
        tools:context=".MainActivity" />
</LinearLayout>
```

在这个文件中可以看到<TextView/>中的属性 text 表示 TextView 要显示的文本，@string/string1 便是文本的内容，但这并不是实际内容而是内容的引用，@string 表示这个内容在字符串资源文件中，string1 是内容在字符串资源文件中的名称。

之后我们再来看在 java 文件中进行引用。

【实例 3.20】如图 3-19 所示，显示字符串资源“你好欢迎再次来到应用系！”。

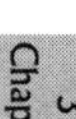

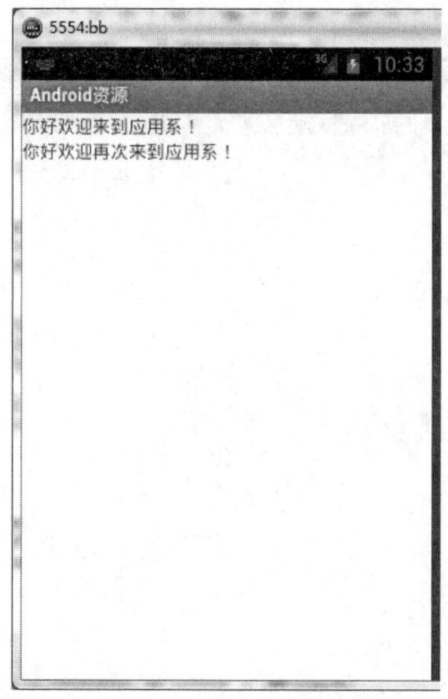

图 3-19　显示字符串资源

步骤一：activity_main.xml 文件中的内容，也就是 activity 的布局文件内容。

```
<LinearLayout xmlns:android="http://schemas.android.com/apk/res/android"
      xmlns:tools="http://schemas.android.com/tools"
      android:layout_width="match_parent"
      android:layout_height="match_parent"
      android:orientation="vertical" >
      <TextView
            android:id="@+id/lblTitle1"
            android:layout_width="fill_parent"
            android:layout_height="wrap_content"
            android:text="@string/string1"
            tools:context=".MainActivity" />
      <TextView
          android:id="@+id/lblTitle2"
          android:layout_width="fill_parent"
          android:layout_height="wrap_content"
          tools:context=".MainActivity" />
</LinearLayout>
```

步骤二：为布局文件添加一个 TextView 组件，但在这个组件中并未进行字符串的引用，字符串的引用被放在了 java 文件中即 MainActivity.java 中，下面便是 MainActivity.java 文件。

```
package com.icer.android_resource;
import android.os.Bundle;
import android.app.Activity;
import android.view.Menu;
import android.widget.TextView;
```

```

public class MainActivity extends Activity {

        // 声明组件
        private TextView lblTitle1,lblTitle2;

    @Override
    public void onCreate(Bundle savedInstanceState) {
        super.onCreate(savedInstanceState);
        setContentView(R.layout.activity_main);

        //获取组件
        this.lblTitle1 = (TextView) findViewById(R.id.lblTitle1);
        this.lblTitle2 = (TextView) findViewById(R.id.lblTitle2);

 //获取字符串
        lblTitle2.setText(this.getResources().getString(R.string.string2));

    }

    @Override
    public boolean onCreateOptionsMenu(Menu menu) {
        getMenuInflater().inflate(R.menu.activity_main, menu);
        return true;
    }
}
```

在 MainActivity.java 文件中，之前新添加的 TextView 组件 lblTitle2 进行字符串引用，其中 setText()意为获取字符串内容，但字符串不能直接获取，我们要以资源的形式获取，getResources()意为获取资源，getString()意为获取字符串，R.string.string2 为想要获取的资源。

3.4.3 颜色资源（Color）

为了让 Android 程序在外观上有更加多彩的表现力，通常会为其中的组件、字体等内容设定各种各样的颜色，这些颜色也被称作颜色资源。为了更加方便地管理和替换这些颜色，Android 系统将这些颜色设定统一记录在 colors.xml 文件中，具体的位置为 res\values\colors.xml。当然如果没有 colors.xml 文件，我们可以自己新建一个，在新建文件选项中选择“Android XML Values File”即可。

colors.xml 文件中的内容：

```
<resources>
    <color name="red">#FF0000</color>
    <color name="blue">#0000FF </color>
</resources>
```

在这个文件中所有的颜色资源要位于<resources></resources>之中，<color name="red">中 name 为颜色的名称属性，red 为这个颜色的名称属性的值，在<color></color>之中的部分即

“#FF0000”为颜色资源的实际内容，也就是颜色的 RGB 代码。

对于颜色资源的引用我们要分为两种情况，第一种是在 xml 文件中进行引用，第二种是在 java 文件中进行引用。

首先我们来看在 xml 文件中进行引用。

【实例 3.21】如图 3-20 所示，文字应用红色颜色资源。

图 3-20　文字颜色设置

步骤一：activity_main.xml 文件中的内容，也就是 activity 的布局文件内容。

```
<LinearLayout xmlns:android="http://schemas.android.com/apk/res/android"
    xmlns:tools="http://schemas.android.com/tools"
    android:layout_width="match_parent"
    android:layout_height="match_parent"
    android:orientation="vertical" >

    <TextView
        android:id="@+id/txtShow1"
        android:layout_width="fill_parent"
        android:layout_height="wrap_content"
        android:textColor="@color/red"
        android:text="@string/hello_world"
        tools:context=".MainActivity" />
</LinearLayout>
<string name="string2">你好欢迎再次来到应用系！</string>
</resources>
```

在这个文件中可以看到<TextView/>中的属性 textColor 表示 TextView 要显示的文本颜色，

@color/red 便是文本颜色的内容，但这只是内容的引用，@color 表示这个内容在颜色资源文件中，red 是内容在颜色资源文件中的名称。

之后我们再来看在 java 文件中进行引用。

【实例 3.22】如图 3-21 所示，文字应用蓝色颜色资源。

图 3-21　文字颜色设置“蓝色”

步骤一：activity_main.xml 文件中的内容，也就是 activity 的布局文件内容。

```
<LinearLayout xmlns:android="http://schemas.android.com/apk/res/android"
xmlns:tools="http://schemas.android.com/tools"
    android:layout_width="match_parent"
    android:layout_height="match_parent"
    android:orientation="vertical" >
<TextView
    android:id="@+id/txtShow1"
    android:layout_width="fill_parent"
    android:layout_height="wrap_content"
      android:text="@string/hello_world"
    tools:context=".MainActivity" />
</LinearLayout>
```

步骤二：在布局文件中并未添加颜色资源，资源被添加在了 java 文件中，即 MainActivity.java。下面便是 MainActivity.java 文件。

```
package com.icer.android_res_2;
import android.os.Bundle;
import android.app.Activity;
import android.view.Menu;
```

```
import android.widget.TextView;
public class MainActivity extends Activity {
    // 声明组件
    private TextView txtShow1;
    @Override
    public void onCreate(Bundle savedInstanceState) {
        super.onCreate(savedInstanceState);
        setContentView(R.layout.activity_main);
      // 获取组件
        this.txtShow1 = (TextView) findViewById(R.id.txtShow1);
Resources res = this.getResources();
      // 设置颜色
        this.txtShow1.setTextColor(res.getColor(R.color.blue));
  }
    @Override
    public boolean onCreateOptionsMenu(Menu menu) {
        getMenuInflater().inflate(R.menu.activity_main, menu);
        return true;
    }
}
```

在 MainActivity.java 文件中，TextView 组件 txtShow1 进行颜色资源的引用，其中 setTextColor()意为获取文本颜色，但颜色不能直接获取，我们要以资源的形式获取，getResources()意为获取资源，getColor()意为获取颜色，R.color.blue 为想要获取的资源。

3.4.4 数组资源（Array）

我们都知道对于相同数据类型的元素使用数组是最佳的选择，这一点到了 Android 中也没有改变，Android 同样引入了数组这一概念，并将其归类为资源供用户使用，具体的位置为 res\values\arrays.xml。当然如果没有 arrays.xml 文件，我们可以自己新建一个，在新建文件选项中选择“Android XML Values File”即可，下面我们就来看一下数组资源的具体内容。

arrays.xml 文件中的内容

```
<resources>
    <array name = "color_array">
        <item>#000000</item>
            <item>#00ff00</item>
    <item>#ff0000</item>
    </array>
    <string-array name = "string_array">
        <item>龙虾</item>
        <item>鲍鱼</item>
        <item>鱼翅</item>
    </string-array>
    <integer-array name = "integer_array">
        <item>1</item>
        <item>2</item>
```

```
            <item>3</item>
        </integer-array>
</resources>
```

在这个文件中所有的数组资源要位于<resources></resources>之中，在这其中我们可以看到处于同一级别的三种不同的标签，即<array>、<string-array>、<integer-array>。它们分别对应不同数组，即<array>对应普通类型数组、<string-array>对应字符串数组、<integer-array>对应整数数组，在这三种类型数组下均有<item></item>，即项作为其子元素，用以存放数组内容。在这个文件中每个数组都拥有自己的属性，以<string-array></string-array>字符串数组为例，它的属性为名称属性 name，其值也就是数组名为"string_array"。

之后我们来看一下布局文件。

【实例 3.23】如图 3-22 所示，字符串数组展示。

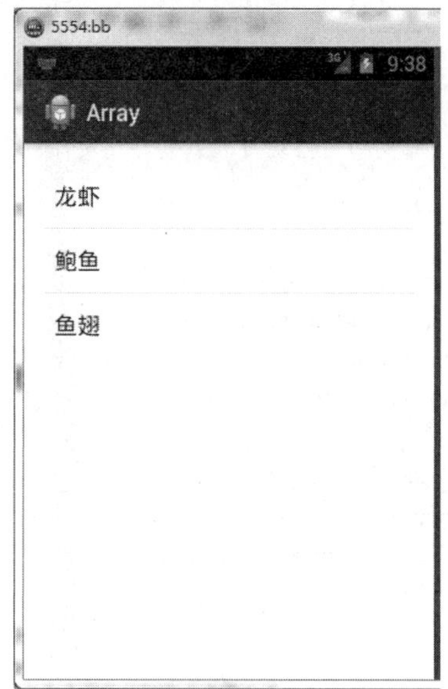

图 3-22　字符串数组设置

步骤一：activity_main.xml 文件中的内容，也就是 activity 的布局文件内容。

```
<RelativeLayout xmlns:android="http://schemas.android.com/apk/res/android"
xmlns:tools="http://schemas.android.com/tools"
android:layout_width="match_parent"
android:layout_height="match_parent"
 android:paddingBottom="@dimen/activity_vertical_margin"
    android:paddingLeft="@dimen/activity_horizontal_margin"
    android:paddingRight="@dimen/activity_horizontal_margin"
    android:paddingTop="@dimen/activity_vertical_margin"
    tools:context=".MainActivity" >
     <ListView
            android:id="@+id/lv"
```

```
            android:layout_width="fill_parent"
            android:layout_height="wrap_content"
            />
</RelativeLayout>
```

在布局文件中我们添加了一个组件<ListView>用于承接数组。

步骤二：数组资源的获取与引用，我们来看一下 java 文件 MainActivity.java，即实现数组资源应用的文件。

```
package com.icer.listviewdemo01;
import android.R.string;
import android.os.Bundle;
import android.app.Activity;
import android.content.res.Resources;
import android.view.Menu;
import android.widget.ArrayAdapter;
import android.widget.ListView;
public class MainActivity extends Activity {
        private ListView lv;
        private Resources res;
        private String[] array_string;
        @Override
        protected void onCreate(Bundle savedInstanceState) {
                super.onCreate(savedInstanceState);
                setContentView(R.layout.activity_main);
                this.lv = (ListView) findViewById(R.id.lv);
                res = getResources();
                array_string = res.getStringArray(R.array.string_array);
                ArrayAdapter<String> adapter = new ArrayAdapter<String>
                        (this,android.R.layout.simple_list_item_1 ,this.array_string);
                this.lv.setAdapter(adapter);
        }
 @Override
      public boolean onCreateOptionsMenu(Menu menu) {
            getMenuInflater().inflate(R.menu.activity_main, menu);
            return true;
      }
}
```

其中，我们通过 ListView lv 声明了一个 ListView，findViewById()意为通过 id 获取组件，通过 getResources()获取资源，通过 res.getStringArray 获取了对应的字符串数组资源，为了要对这个数组进行显示我们要用到适配器，也就是 ArrayAdapter<String> adapter，在这里数据的数据类型为 String 类型，在创建的适配器当中的三个参数分别为“上下文”“android 内置的一个布局方式”“显示的内容”；之后我们要进行一下适配器的绑定，通过语句 this.lv.setAdapter()来实现，这样在运行程序之后便可以得到垂直排列的显示数组资源的结果了。

3.4.5 背景选择器（Selector）

对于使用 Android 开发的程序界面，可在不同状态下设置不同的背景，这是一种动态的变化，例如某一按钮在按下与抬起两个状态时，其所在界面背景是不相同的，但是如果这种动态的变化在代码中进行设置，是相对比较麻烦的，为此 Android 为我们提供了一种解决方案，即背景选择器 selector。

首先我们要在 drawable 文件夹内新建一个.xml 的资源文件，在其中写入选择器的内容，下面我们举一个 xml 文件 bg_button.xml 实例。

```
<?xml version="1.0" encoding="UTF-8"?>
<selector xmlns:android="http://schemas.android.com/apk/res/android">
    <item android:state_checked="true" android:drawable="@drawable/ps1"></item>
    <item android:state_checked="false" android:drawable="@drawable/ps2"></item>
</selector>
```

在实例中我们可以看到在<selector></selector>下存在多个<item></item>项，每一个项代表一种背景。每个项都有自己的属性，而正是这些属性实际完成了背景转换效果，属性 android:state_checked 表示控件勾选，其值为 true 时表示被勾选，值为 false 时表示未被勾选。属性 android:drawable 表示获取图片，通过其值我们也可以发现，两个项所获取的图片并不相同。

在背景选择器设置完成后，我们要对其进行引用。为了比较容易的查看效果，我们使用 CheckBox 组件进行展示。

【实例 3.24】如图 3-23 所示，白猫背景选择器展示。

图 3-23 设置白猫背景选择器设置

步骤一：activity_main.xml 文件

```
<LinearLayout xmlns:android="http://schemas.android.com/apk/res/android"
    xmlns:tools="http://schemas.android.com/tools"
    android:layout_width="match_parent"
    android:layout_height="match_parent"
    android:orientation="vertical" >
    <CheckBox
        android:id="@+id/cb1"
        android:layout_width="wrap_content"
        android:layout_height="wrap_content"
        android:button="@null"
        android:textSize="15sp"
        android:textColor="#EE2C2C"
        android:drawableTop="@drawable/bg_button"
        android:text="多选 1"
        android:checked="true"/>
    <CheckBox
        android:id="@+id/cb2"
        android:layout_width="wrap_content"
        android:layout_height="wrap_content"
        android:button="@null"
        android:textSize="15sp"
        android:textColor="#EE2C2C"
        android:drawableTop="@drawable/bg_button”
        android:text="多选 2"/>
</LinearLayout>
```

在这我们可以看到在<CheckBox></CheckBox>中对于背景选择器的引用是通过属性android:drawableTop 来完成的，也就是说其实背景选择器被当做图片资源所使用的。

背景选择器用于实现背景转换的属性并不是只有我们在实例中给大家展示的那一种，它所具有的属性是多种多样的，例如：

android:state_enabled="true/false"意为是否能够接受触摸或者点击事件；

android:state_focused="true/false"意为是否取得焦点；

android:state_pressed="true/false"意为是否按下（按钮等）；

android:state_selected="true/false"意为是否被选中。

3.5 单选按钮和相关事件

Android 提供了单项选择的组件，可通过 RadioGroup、RadioButton 组合起来完成一个单选选择效果。一个单项选择由两部分组成，分别是前面的选择按钮和后面的“答案”。Android 平台上的选择按钮可以通过 RadioButton 来实现，而“答案”则通过 RadioGroup 来实现。因此，首先要在布局文件中定义一个 RadioGroup 和若干 RadioButton，在定义 RadioGroup 时，已经将“答案”赋给了每个选项，那么如何确定用户的选择是否正确呢？这需要在用户点击时来判断用户选择的是哪一项，所以需要设置事件监听 setOnCheckedChangeListener。通过下面实例，了解单选按钮和 OnCheckedChangeListener 选择事件。

【实例 3.25】完成如图 3-24 所示的“计算机试题测试”，当用户选择答案不正确时，显示图 3-24（b）所示信息，当用户选择答案正确时，显示图 3-24（c）所示信息。

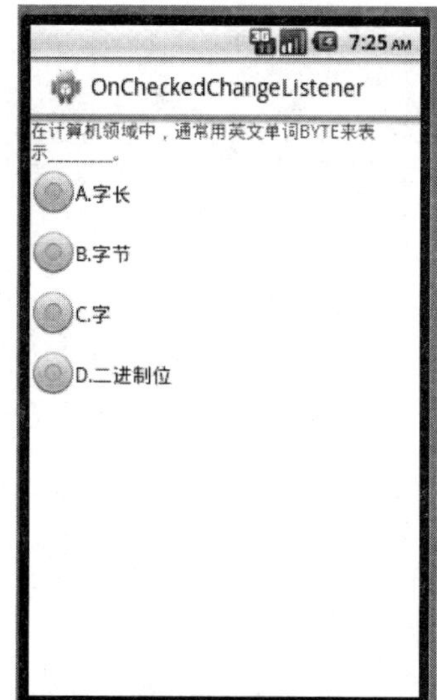

（a）试题界面

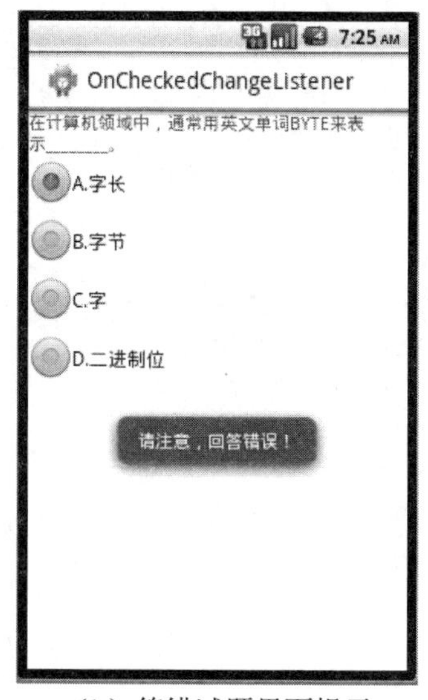

（b）答错试题界面提示

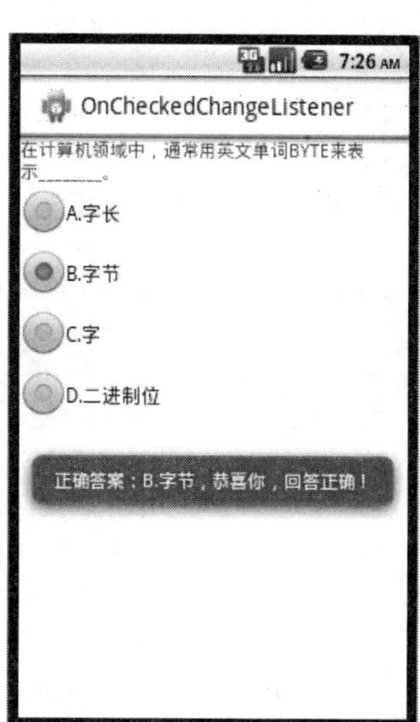

（c）答对试题界面提示

图 3-24　单选按钮实例

步骤一：创建名为“RadioGroupExample”的工程，在 res/layout/activity_main.xml 中添加如下代码：

```
<LinearLayout xmlns:android="http://schemas.android.com/apk/res/android"
        android:layout_width="match_parent"
        android:layout_height="match_parent"
        android:orientation="vertical" >
    <TextView
        android:id="@+id/tv1"
        android:layout_width="fill_parent"
        android:layout_height="wrap_content"
        android:text="@string/hello" />
    <RadioGroup
        android1:id="@+id/radioGroup1"
        android1:layout_width="wrap_content"
        android1:layout_height="wrap_content"
```

```
            android:orientation="vertical"
            android:layout_x="3px"
            android:layout_y="54px" >
        <RadioButton
                android1:id="@+id/radio0"
                android1:layout_width="wrap_content"
                android1:layout_height="wrap_content"
                android1:text="@string/radio0" />
          <RadioButton
                android1:id="@+id/radio1"
                android1:layout_width="wrap_content"
                android1:layout_height="wrap_content"
                android1:text="@string/radio1" />
          <RadioButton
                android1:id="@+id/radio2"
                android1:layout_width="wrap_content"
                android1:layout_height="wrap_content"
                android1:text="@string/radio2" />
           <RadioButton
                android1:id="@+id/radio3"
                android1:layout_width="wrap_content"
                android1:layout_height="wrap_content"
                android1:text="@string/radio3" />
         </RadioGroup>
</LinearLayout>
```

步骤二：在 res/values/strings.xml 中，修改以下代码。

```
<?xml version="1.0" encoding="utf-8"?>
<resources>
     <string name="app_name">OnCheckedChangeListener</string>
       <string name="action_settings">Settings</string>
       <string name="radio0">A.字长</string>
       <string name="radio1">B.字节</string>
       <string name="radio2">C.字</string>
<string name="radio3">D.二进制位</string>
<string name="hello">
在计算机领域中，通常用英文单词"BYTE"来表示________。</string>
</resources>
```

步骤三：在 src 目录下打开 MainActivity.java 文件，修改文件，内容如下：

```
public class MainActivity extends ActionBarActivity {
<!—创建 TextView、RadioGroup、4 个 RadioButton 对象-->
      TextView m_TextView;
      RadioGroup m_radioGroup;
      RadioButton ad1,ad2,ad3,ad4;
```

```
    @Override
    protected void onCreate(Bundle savedInstanceState) {
        super.onCreate(savedInstanceState);
        setContentView(R.layout.activity_main);
<!—获得 TextView、RadioGroup、4 个 RadioButton 对象-->
        m_TextView=(TextView)findViewById(R.id.tv1);
        m_radioGroup=(RadioGroup)findViewById(R.id.radioGroup1);
        ad1=(RadioButton)findViewById(R.id.radio0);
        ad2=(RadioButton)findViewById(R.id.radio1);
        ad3=(RadioButton)findViewById(R.id.radio2);
        ad4=(RadioButton)findViewById(R.id.radio3);
    <!—设置事件监听-->
m_radioGroup.setOnCheckedChangeListener(new RadioGroup.OnCheckedChangeListener(){
    public void onCheckedChanged(RadioGroup group,int checkedId)
            {
                if(checkedId==ad2.getId())
                    {
                        DisplayToast("正确答案："+ad2.getText()+"，恭喜你，回答正确！");
                }
                else
                {
                        DisplayToast("请注意，回答错误！");
                }
            }
        });
    }
<!—显示 Toast-->
public void DisplayToast(String str){
    Toast toast=Toast.makeText(this, str,Toast.LENGTH_LONG);
    toast.setGravity(Gravity.TOP,0, 220);
    toast.show();
    }
}
```

步骤四：运行项目。

3.6　多项选择和相关事件

多项选择和单项选择最重要的区别在于可以让用户选择一个以上的选项。Android 平台提供了 CheckBox 来实现多选。既然用户可以选择多项，那么为了确定用户是否选择了某一项，需要对每一个选项进行事件监听。

【实例 3.26】完成如图 3-25 所示的“计算机试题测试程序”，当用户选择完毕并提交之后，需要给用户反馈信息，如图 3-25（c）所示。

（a）试题界面

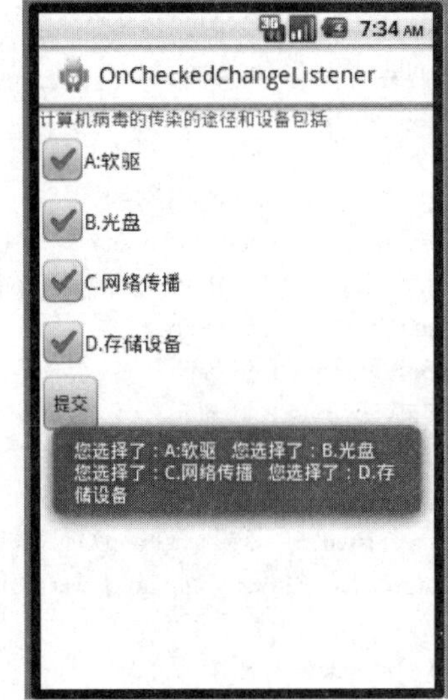

（b）单击试题答案后界面提示

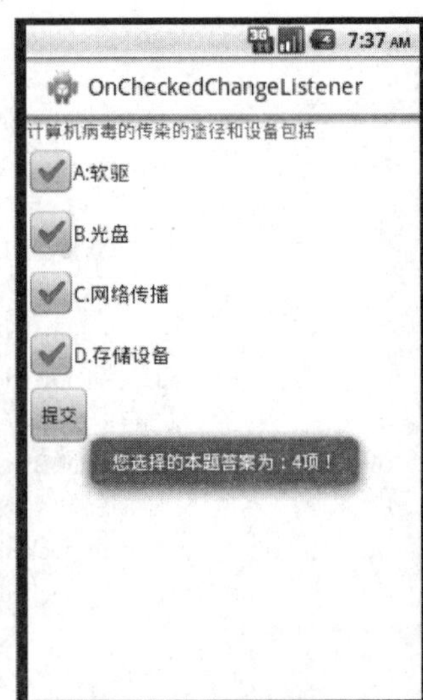

（c）提交后界面提示

图 3-25 多项选择实例

步骤一：创建名为“checkExample”的工程，在 res/layout/activity_main.xml 中添加如下代码：

```
<LinearLayout xmlns:android="http://schemas.android.com/apk/res/android"
xmlns:android1="http://schemas.android.com/apk/res/android"
        android:layout_width="match_parent"
        android:layout_height="match_parent"
        android:orientation="vertical" >
        <TextView
            android:id="@+id/tv1"
            android:layout_width="fill_parent"
            android:layout_height="wrap_content"
            android:text="@string/hello" />
        <CheckBox
            android1:id="@+id/ck1"
            android1:layout_width="fill_parent"
            android1:layout_height="wrap_content"
            android1:text="@string/ck1" >
    </CheckBox>
        <CheckBox
            android1:id="@+id/ck2"
            android1:layout_width="fill_parent"
            android1:layout_height="wrap_content"
            android1:text="@string/ck2" >
    </CheckBox>
        <CheckBox
            android1:id="@+id/ck3"
```

```
            android1:layout_width="fill_parent"
            android1:layout_height="wrap_content"
            android1:text="@string/ck3" >
</CheckBox>
        <CheckBox
            android1:id="@+id/ck4"
            android1:layout_width="fill_parent"
            android1:layout_height="wrap_content"
            android1:text="@string/ck4" >
</CheckBox>
        <Button
            android1:id="@+id/button1"
            android1:layout_width="wrap_content"
            android1:layout_height="wrap_content"
            android1:text="提交" />
</LinearLayout>
```

步骤二：在 res/values/strings.xml 中，修改以下代码：

```
<?xml version="1.0" encoding="utf-8"?>
<resources>
        <string name="app_name">OnCheckedChangeListener</string>
        <string name="action_settings">Settings</string>
        <string name="ck1">A:键盘</string>
        <string name="ck2">B.字节</string>
        <string name="ck3">C.字</string>
<string name="ck4">D.二进制位</string>
<string name="hello">计算机病毒的传染的途径和设备包括</string>
</resources>
```

步骤三：在 src 目录下打开 MainActivity.java 文件，修改文件，内容如下：

```
public class MainActivity extends ActionBarActivity {
        TextView m_TextView;
        Button Button1;
        CheckBox cb1,cb2,cb3,cb4;
        String str="";
        @Override
        protected void onCreate(Bundle savedInstanceState) {
                super.onCreate(savedInstanceState);
                setContentView(R.layout.activity_main);
                m_TextView=(TextView)findViewById(R.id.tv1);
                Button1=(Button)findViewById(R.id.button1);
                cb1=(CheckBox)findViewById(R.id.ck1);
                cb2=(CheckBox)findViewById(R.id.ck2);
                cb3=(CheckBox)findViewById(R.id.ck3);
                cb4=(CheckBox)findViewById(R.id.ck4);
                cb1.setOnCheckedChangeListener(new CheckBox.OnCheckedChangeListener(){
                        @Override
                        public void onCheckedChanged(CompoundButton buttonView,boolean ischecked)
                        {
```

```
                        if(cb1.isChecked())
                        {   str=str+"您选择了："+cb1.getText()+"    ";
                              DisplayToast(str);
                              }
                  }
            });
            cb2.setOnCheckedChangeListener(new CheckBox.OnCheckedChangeListener(){
                  public void onCheckedChanged(CompoundButton buttonView,boolean ischecked)
                  {
                        if(cb2.isChecked())
                        {str=str+"您选择了："+cb2.getText()+"    ";
                              DisplayToast(str);
                        }
                  }
            });
            cb3.setOnCheckedChangeListener(new CheckBox.OnCheckedChangeListener(){
                  public void onCheckedChanged(CompoundButton buttonView,boolean ischecked)
                  {
                        if(cb3.isChecked())
                        {str=str+"您选择了："+cb3.getText()+"    ";
                              DisplayToast(str);
                        }
                  }
            });
            cb4.setOnCheckedChangeListener(new CheckBox.OnCheckedChangeListener(){
                  public void onCheckedChanged(CompoundButton buttonView,boolean ischecked)
                  {
                        if(cb4.isChecked())
                        {str=str+"您选择了："+cb4.getText()+"    ";
                              DisplayToast(str);
                        }
                  }
            });
            Button1.setOnClickListener(new Button.OnClickListener(){
                  public void onClick(View v){
                        int num=0;
                        if(cb1.isChecked())
                              num++;
                        if(cb2.isChecked())
                              num++;
                        if(cb3.isChecked())
                              num++;
                        if(cb4.isChecked())
                              num++;
                        DisplayToast("您选择的本题答案为："+num+"项！");
                  }
            });
      }
```

```
public void DisplayToast(String str){
        Toast toast=Toast.makeText(this, str,Toast.LENGTH_SHORT);
        toast.setGravity(Gravity.TOP,0, 220);
        toast.show();
        }
}
```

步骤四：运行项目。

3.7 实训项目

3.7.1 开发标准身高计算器

身高体重指数（又称身体质量指数，英文为 Body Mass Index，简称 BMI）是一个计算值。用来比较及分析一个人的体重对于不同高度的人所带来的健康影响。本程序需要输入身高、体重，按下“计算 BMI 值”键后就在屏幕上显示 BMI 值，并弹出“你应该节食”“你应该多吃点”…等健康建议。健康建议的判断：只要 BMI 值超过 25 时就算偏胖、BMI 值低于 20 就算偏瘦，如图 3-26 所示。

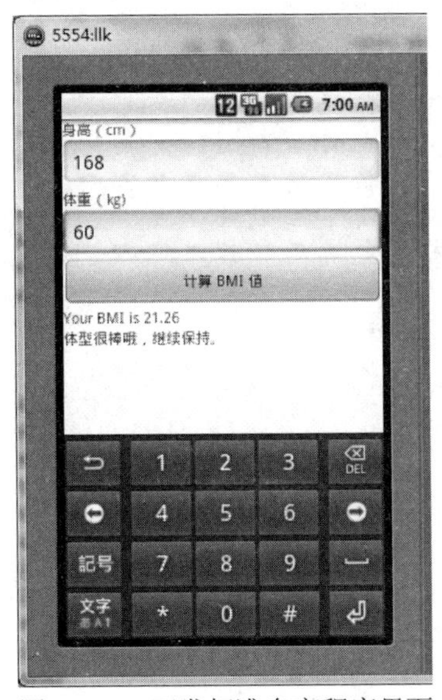

图 3-26　开发标准身高程序界面

步骤一：创建名为“Bmi”的工程，在 res/layout/activity_main.xml 中添加如下代码：

```
<?xml version="1.0" encoding="utf-8"?>
<LinearLayout xmlns:android="http://schemas.android.com/apk/res/android"
        android:orientation="vertical"
        android:layout_width="fill_parent"
```

```
                android:layout_height="fill_parent"    >
    <TextView
                android:layout_width="fill_parent"
                android:layout_height="wrap_content"
                android:text="@string/height"
        />
    <EditText android:id="@+id/height"
                android:layout_width="fill_parent"
                android:layout_height="wrap_content"
                android:numeric="integer"
                android:text=""
        />
    <TextView
                android:layout_width="fill_parent"
                android:layout_height="wrap_content"
                android:text="@string/weight"
        />
    <EditText android:id="@+id/weight"
                android:layout_width="fill_parent"
                android:layout_height="wrap_content"
                android:numeric="integer"
                android:text=""
        />
    <Button android:id="@+id/submit"
              android:layout_width="fill_parent"
              android:layout_height="wrap_content"
              android:text="@string/bmi_btn"
        />
    <TextView    android:id="@+id/result"
              android:layout_width="fill_parent"
              android:layout_height="wrap_content"
              android:text=""
        />
    <TextView    android:id="@+id/suggest"
              android:layout_width="fill_parent"
              android:layout_height="wrap_content"
              android:text=""
        />
    </LinearLayout>
```

步骤二：在 res/values/中，新建 advice.xml，代码如下所示：

```
<resources>
    <string name="advice_light">你该多吃点，身体是革命的本钱哈。</string>
    <string name="advice_average">体型很棒哦，继续保持。</string>
    <string name="advice_heavy">你该节食了，呵呵。</string>
</resources>
```

步骤三：修改 value 中的 string.xml

```
    <resources>
```

```
    <string name="app_name">BIM</string>
    <string name="height">身高（cm）</string>
    <string name="weight">体重（kg)</string>
    <string name="bmi_btn">计算 BMI 值</string>
    <string name="bmi_result">您的 BIM 值是：</string>
</resources>
```

步骤四：在 src 目录下打开 MainActivity.java 文件，修改文件，内容如下：

```
public class MainActivity extends Activity {
    @Override
    public void onCreate(Bundle savedInstanceState) {
        super.onCreate(savedInstanceState);
        setContentView(R.layout.activity_main);
        // Listen for button clicks
        Button button = (Button) findViewById(R.id.submit);
        button.setOnClickListener(calcBMI);
    }
    private OnClickListener calcBMI = new OnClickListener() {
        public void onClick(View v) {
            DecimalFormat nf = new DecimalFormat("0.00");
    EditText fieldheight = (EditText) findViewById(R.id.height);
    EditText fieldweight = (EditText) findViewById(R.id.weight);
            double height = Double
                    .parseDouble(fieldheight.getText().toString()) / 100;
            double weight = Double
                    .parseDouble(fieldweight.getText().toString());
            double BMI = weight / (height * height);
            TextView result = (TextView) findViewById(R.id.result);
            result.setText("Your BMI is " + nf.format(BMI));
            TextView fieldsuggest = (TextView) findViewById(R.id.suggest);
            if (BMI > 25) {
                fieldsuggest.setText(R.string.advice_heavy);
            } else if (BMI < 20) {
                fieldsuggest.setText(R.string.advice_light);
            } else {
                fieldsuggest.setText(R.string.advice_average);
            }
        }
    };
}
```

步骤五：运行项目。

3.7.2 制作手机桌面

手机桌面是手机的门户，通常手机桌面会放置系统的应用程序及应用服务等。应用程序包含图标和名称。实现手机桌面应用，需要实现以下三个功能：

- 应用图标的排放布局。

- 手机图标要求“见标知意”（见到图标，知道应用程序的功能）。
- 监听器管理，当单击或触摸时可以立即打开相关的应用程序。

【实例 3.27】制作手机桌面如图 3-27 所示。

图 3-27　手机界面

步骤一：创建名为“desklayout”的工程，在 res/layout/activity_main.xml 中添加如下代码：

```
<LinearLayout xmlns:android="http://schemas.android.com/apk/res/android"
      android:layout_width="fill_parent"
      android:layout_height="fill_parent"
      android:orientation="vertical" >
<LinearLayout
         android:layout_width="fill_parent"
         android:layout_height="fill_parent"
         android:id="@+id/linearLayout1"
         android:layout_weight="1"
         android:orientation="horizontal"
         android:layout_marginTop="40px">
      <TextView
          android:layout_width="20px"
          android:layout_height="40px" />
      <LinearLayout
         android:layout_width="wrap_content"
         android:layout_height="70px"
         android:orientation="vertical" >
      <ImageButton
         android:id="@+id/imageButton11"
         android:layout_width="40px"
```

```
            android:layout_height="40px"
            android:focusableInTouchMode="true"
            android:src="@drawable/icon1" />
         <TextView
            android:layout_width="wrap_content"
            android:layout_height="20px"
            android:text="图片"/>
   </LinearLayout>
        <TextView
            android:layout_width="40px"
            android:layout_height="40px" />
        <LinearLayout
            android:layout_width="wrap_content"
            android:layout_height="70px"
            android:orientation="vertical" >
        <ImageButton
            android:id="@+id/imageButton12"
            android:layout_width="40px"
            android:layout_height="40px"
            android:focusableInTouchMode="true"
            android:src="@drawable/icon2" />
         <TextView
            android:layout_width="wrap_content"
            android:layout_height="20px"
            android:text="编辑"/>
        </LinearLayout>
        <TextView
            android:layout_width="40px"
            android:layout_height="40px"/>
        <LinearLayout
            android:layout_width="wrap_content"
            android:layout_height="70px"
            android:orientation="vertical" >
        <ImageButton
            android:id="@+id/imageButton13"
            android:layout_width="40px"
            android:layout_height="40px"
            android:focusableInTouchMode="true"
            android:src="@drawable/icon3" />
         <TextView
            android:layout_width="wrap_content"
            android:layout_height="20px"
            android:text="测试"/>
        </LinearLayout>
   </LinearLayout>
        <!--桌面第二行了  -->
           <LinearLayout
               android:layout_width="fill_parent"
```

```
                android:layout_height="fill_parent"
                android:id="@+id/linearLayout1"
                android:orientation="horizontal"
                android:layout_weight="1">
              <TextView
                  android:layout_width="20px"
                  android:layout_height="40px"/>
        <LinearLayout
              android:layout_width="wrap_content"
              android:layout_height="70px"
              android:orientation="vertical" >
        <ImageButton
              android:id="@+id/imageButton21"
              android:layout_width="40px"
              android:layout_height="40px"
              android:focusableInTouchMode="true"
              android:src="@drawable/icon4" />
         <TextView
                  android:layout_width="wrap_content"
                  android:layout_height="20px"
                  android:text="录音"
                  />
    </LinearLayout>
        <TextView
                  android:layout_width="40px"
                  android:layout_height="40px"/>
        <LinearLayout
              android:layout_width="wrap_content"
              android:layout_height="70px"
              android:orientation="vertical" >
        <ImageButton
              android:id="@+id/imageButton22"
              android:layout_width="40px"
              android:layout_height="40px"
              android:focusableInTouchMode="true"
              android:src="@drawable/icon5" />
         <TextView
                  android:layout_width="wrap_content"
                  android:layout_height="20px"
                  android:text="诱饵"
                  />
    </LinearLayout>
        <TextView
                  android:layout_width="40px"
                  android:layout_height="40px"
                  />
        <LinearLayout
              android:layout_width="wrap_content"
```

```
                android:layout_height="70px"
                android:orientation="vertical" >
            <ImageButton
                android:id="@+id/imageButton23"
                android:layout_width="40px"
                android:layout_height="40px"
                android:focusableInTouchMode="true"
                android:src="@drawable/icon6" />
             <TextView
                android:layout_width="wrap_content"
                android:layout_height="20px"
                android:text="自传"
                    />
      </LinearLayout>
    </LinearLayout>
    </LinearLayout>
```

步骤二：在 src 目录下打开 MainActivity.java 文件，修改文件，内容如下：

```
public class MainActivity extends ActionBarActivity {
        private ImageButton IB_app1;
        private ImageButton IB_app2;
        private ImageButton IB_app3;
        private ImageButton IB_app4;
        private ImageButton IB_app5;
        private ImageButton IB_app6;
        private Intent Intent_app1;
        private Intent Intent_app2;
        private Intent Intent_app3;
        private Intent Intent_app4;
        private Intent Intent_app5;
        private Intent Intent_app6;
        public void onCreate(Bundle savedInstanceState)
        {try
        {
                super.onCreate(savedInstanceState);
                init();
                setContentView(R.layout.activity_main);
                createComponent();
                register();
        }
        catch(Exception e)
        {
                Toast.makeText(MainActivity.this,"异常错误："+e.toString(),Toast.LENGTH_LONG).show();
        }
        finally{}
}
private void init()
{
```

```
            IB_app1=null;
            IB_app2=null;
            IB_app3=null;
            IB_app4=null;
            IB_app5=null;
            IB_app6=null;
            Intent_app1=null;
            Intent_app2=null;
            Intent_app3=null;
            Intent_app4=null;
            Intent_app5=null;
            Intent_app6=null;}
    private void createComponent(){
    IB_app1=(ImageButton)this.findViewById(R.id.imageButton11);
    IB_app2=(ImageButton)this.findViewById(R.id.imageButton12);
    IB_app3=(ImageButton)this.findViewById(R.id.imageButton13);
    IB_app4=(ImageButton)this.findViewById(R.id.imageButton21);
    IB_app5=(ImageButton)this.findViewById(R.id.imageButton22);
    IB_app6=(ImageButton)this.findViewById(R.id.imageButton23);
            }
    private void register(){
            IB_app1.setOnClickListener(new OnClickListener(){
                    public void onClick(View v){
                            Intent_app1=new Intent(MainActivity.this,MainActivity.class);
                            startActivity(Intent_app1);
                    }
            });
            IB_app2.setOnClickListener(new OnClickListener(){
                    public void onClick(View v){
                            Intent_app2=new Intent(MainActivity.this,MainActivity.class);
                            startActivity(Intent_app2);
                    }
            });
            IB_app3.setOnClickListener(new OnClickListener(){
                    public void onClick(View v){
                            Intent_app3=new Intent(MainActivity.this,MainActivity.class);
                            startActivity(Intent_app3);
                    }
            });
            IB_app4.setOnClickListener(new OnClickListener(){
                    public void onClick(View v){
                            Intent_app4=new Intent(MainActivity.this,MainActivity.class);
                            startActivity(Intent_app4);
                    }
            });
            IB_app5.setOnClickListener(new OnClickListener(){
                    public void onClick(View v){
                            Intent_app5=new Intent(MainActivity.this,MainActivity.class);
```

```
                    startActivity(Intent_app5);
                }
        });
        IB_app6.setOnClickListener(new OnClickListener(){
                public void onClick(View v){
                    Intent_app6=new Intent(MainActivity.this,MainActivity.class);
                    startActivity(Intent_app6);
                }
        });
}
```

步骤三：运行项目。

3.7.3　调查问卷程序

【实例 3.28】制作调查问卷程序，如图 3-28 所示。

图 3-28　调查问卷程序界面

步骤一：创建名为“asksystemExample”的工程，在 res/layout/activity_main.xml 中添加如下代码：

```
<?xml version="1.0" encoding="utf-8"?>
<LinearLayout xmlns:android="http://schemas.android.com/apk/res/android"
    android:layout_width="fill_parent"
    android:layout_height="wrap_content"
    android:orientation="vertical" >
     <TextView
            android:id="@+id/textView1"
            android:layout_width="wrap_content"
            android:layout_height="wrap_content"
```

```
                android:text="@string/sex"
                android:textSize="10px" />
        <RadioGroup
            android:id="@+id/sexmenu"
            android:layout_width="fill_parent"
            android:layout_height="wrap_content"
            android:orientation="horizontal"
            android:checkedButton="@+id/man">
        <RadioButton
            android:id="@+id/man"
            android:layout_width="wrap_content"
            android:layout_height="wrap_content"
            android:text="@string/man" />
        <RadioButton
           android:id="@+id/woman"
            android:layout_width="wrap_content"
            android:layout_height="wrap_content"
            android:text="@string/woman" />
        </RadioGroup>
        <TextView
            android:id="@+id/textView2"
            android:layout_width="wrap_content"
            android:layout_height="wrap_content"
            android:text="@string/major"
            android:textSize="10px" />
        <CheckBox android:id="@+id/jsj"
            android:layout_width="wrap_content"
            android:layout_height="wrap_content"
            android:text="@string/jsj"/>
        <CheckBox android:id="@+id/xxgl"
            android:layout_width="wrap_content"
            android:layout_height="wrap_content"
            android:text="@string/xxgl"/>
        <CheckBox android:id="@+id/ydhl"
           android:layout_width="wrap_content"
            android:layout_height="wrap_content"
            android:text="@string/ydhl"/>
      <LinearLayout
            android:layout_width="fill_parent"
            android:layout_height="fill_parent"
            android:orientation="horizontal" >
      <TextView
            android:id="@+id/textView3"
            android:layout_width="wrap_content"
            android:layout_height="wrap_content"
            android:text="@string/like"
            android:textSize="10px" />
```

```
        <CheckBox android:id="@+id/basketball"
                android:layout_width="wrap_content"
                android:layout_height="wrap_content"
                android:text="@string/basketball"/>
        <CheckBox
                android:id="@+id/football"
                android:layout_width="wrap_content"
                android:layout_height="wrap_content"
                android:text="@string/football" />
        <CheckBox android:id="@+id/volleyball"
                android:layout_width="wrap_content"
                android:layout_height="wrap_content"
                android:text="@string/volleyball"/>
    </LinearLayout>
<LinearLayout
        android:layout_width="fill_parent"
        android:layout_height="fill_parent"
        android:orientation="vertical" >
        <Button android:id="@+id/editButton"
            android:text="点击提交"
            android:layout_width="fill_parent"
            android:layout_height="wrap_content"
            android:layout_gravity="center_horizontal"
        />
        <TextView
                android:id="@+id/edittext"
                android:layout_width="fill_parent"
                android:layout_height="wrap_content"
                android:textSize="10px" />
        </LinearLayout>
</LinearLayout>
```

步骤二：在 res/values/strings.xml 中，修改以下代码。

```
<resources>
        <string name="app_name">问卷系统</string>
      <string name="sex">性别</string>
        <string name="like">爱好</string>
        <string name="major">专业</string>
        <string name="man">男</string>
        <string name="woman">女</string>
        <string name="basketball">篮球</string>
        <string name="football">足球</string>
        <string name="volleyball">排球</string>
        <string name="jsj">计算机应用专业</string>
        <string name="xxgl">信息管理专业</string>
        <string name="ydhl">移动互联专业</string>
</resources>
```

步骤三：在 src 目录下打开 MainActivity.java 文件，修改文件，内容如下：

```
public class MainActivity extends ActionBarActivity implements OnClickListener{
            RadioButton rbman=null;
            RadioButton rbwoman=null;
            CheckBox cbfoot=null;
            CheckBox cbbasket=null;
            CheckBox cbvolley=null;
            CheckBox cbjsj=null;
            CheckBox cbxxgl=null;
            CheckBox cbydhl=null;
            Button btn=null;
            TextView tView=null;
            @Override
      public void onCreate(Bundle savedInstanceState) {
            super.onCreate(savedInstanceState);
            setContentView(R.layout.activity_main);
          rbman=(RadioButton)findViewById(R.id.man);
          rbwoman=(RadioButton)findViewById(R.id.woman);
          cbbasket=(CheckBox)findViewById(R.id.basketball);
          cbfoot=(CheckBox)findViewById(R.id.football);
          cbvolley=(CheckBox)findViewById(R.id.volleyball);
          cbjsj=(CheckBox)findViewById(R.id.jsj);
          cbxxgl=(CheckBox)findViewById(R.id.xxgl);
          cbydhl=(CheckBox)findViewById(R.id.ydhl);
          btn=(Button)findViewById(R.id.editButton);
          tView=(TextView)findViewById(R.id.edittext);
          btn.setOnClickListener(this);}
          public void onClick(View v) {
     int num=0;
     if(v==btn)
      {tView.setText("你的性别为：");
if(rbman.isChecked())
       tView.append("男\n");
if(rbwoman.isChecked())
      tView.append(rbwoman.getText().toString()+"\n");
if(cbjsj.isChecked())
{tView.append("您的专业是计算机应用\n");}
if(cbxxgl.isChecked())
{tView.append("您的专业是计算机信息管理\n");}
if(cbydhl.isChecked())
{tView.append("您的专业是移动互联\n");}
if(cbbasket.isChecked())
          {num++;tView.append(" 篮球");}
if(cbfoot.isChecked())
{num++;tView.append(" 足球");}
```

```
        if(cbvolley.isChecked())
        {num++;tView.append(" 排球");}
        if(num>0){tView.append("\n 您的爱好有"+num+"种，谢谢您的参与！");}
    }
  }
}
```

步骤四： 运行项目。

3.8　本章小结

本章对 Android 系统一些常用的组件及界面布局进行了详细的介绍，并列举简单示例。通过示例代码可以让学习者加深对各组件功能的认识，让大家综合运用这些组件来完成 Android 开发的 UI 设计。这里的很多组件都能实现与用户进行交互的功能，因此要对需要交互的组件设置事件监听，捕捉用户所触发的事件，从而进行相应的处理。

通过本章的学习，能熟练使用常用组件进行编程。

3.9　本章习题

1．Android 中共有几种常用的文本组件，请描述其作用和区别？
2．ListView 和 ListActivity 有什么联系？
3．请使用单选和多选组件创建“心理测试”，界面自行设计。
4．Android 应用中常用哪些布局管理器，请列举说明布局效果。
5．创建一个对话框，对话框中显示单选按钮列表，内容包括不同网站的链接。
6．设计菜单，包括子菜单设计，内容自拟。

4

Andriod 消息与广播

学习目标：

了解 Android 中的 Intent 组件，并学会灵活使用这组件完成项目的开发。学习 BroadcastReceiver 和 Intent 的组合应用。

【知识目标】

- Intent 界面跳转及传值
- 显式和隐式 Intent 使用
- 接收广播消息
- 发送广播消息

【技能目标】

- 熟练掌握 Intent 常用的几种用法
- 熟练掌握使用 BroadcastReceiver 组件发送和接收广播的方法

之前的内容曾多次使用 Intent 对象进行 Activity 跳转。Intent 除此之外还有更多功能，它可以携带数据、发送广播和启动服务。因此，也可以认为 Intent 是一种对操作的抽象，这些操作包括显示 Activity、传值、发送广播和启动服务。下面将详细介绍 Intent 更多的使用方法。

4.1 Intent 与 Activity

Intent（意图）主要解决 Android 应用的各项组件之间的通讯和发送系统级的广播。下面详细介绍 Android 系统的组件通信原理，掌握利用组件通信传值、启动其他组件和发送广播消息的方法。

4.1.1 Intent 简介

Intent 是一个动作的完整描述，包含了动作的产生组件、接收组件和传递的数据信息。当然，也可以将 Intent 视为不同组件之间传递的消息，这个消息在到达接收组件后，接收组件会执行相关的动作。

由于 Intent 的存在，使得 Android 系统中相互独立的应用程序组件，成为了一个可以互相通信的组件集合。因此，无论这些组件是否在同一个应用程序中，Intent 可以将一个组件的数据和动作传递给另一个组件。

Intent 为 Activity、Service 和 BroadcastReceiver 等组件提供交互能力。Intent 的一个最常见的用途就是启动 Activity 和 Service，另一个用途是在 Android 系统上发布广播消息。广播消息可以是接收的特定数据或者消息，也可以是手机的信号变化或电池的电量过低等信息。

Intent 是由组件名称、Action、Data、Category、Extra 及 Flag 六部分组成的。接下来将分别对其进行详细介绍。

组件名称实际上就是一个 ComponentName 对象，用于标识唯一的应用程序组件，即指明了期望的 Intent 组件，这种对象的名称是由目标组件的类名和同目标组件的包名组合而成的。在 Intent 传递过程中，组件名称是一个可选项，当指定它时，便是显式的 Intent 消息，当不指定它时，Android 系统则会根据其他信息及 IntentFilter 的过滤条件选择相应的组件。

Action 实际上就是一个描述了 Intent 所触发动作名称的字符串， 在 Intent 类中，已经定义好很多字符串常量来表示不同的 Action。当然，开发人员也可以自定义 Action，其定义规则同样非常简单。

系统定义的 Action 常量有很多，下面只列出其中一些较常用的以供参考：

- ACTION_CALL：拨出 Data 里封装的电话号码；
- ACTION_EDIT：打开 Data 里指定数据所对应的编辑程序；
- ACTION_VIEW：打开能够显示 Data 中封装的数据的应用程序；
- ACTION_MAIN：声明程序的入口，该 Action 并不会接收任何数据，同时结束后也不会返回任何数据；
- ACTION_BOOT_COMPLETED：BroadcastReceiverAction 的常量，表明系统启动完毕；
- ACTION_TIME_CHANGED：BroadcastReceiverAction 的常量，表明系统时间通过设置而改变。

Data 主要是对 Intent 消息中数据的封装，主要描述 Intent 的动作所操作数据的 URI 及其类型。不同类型的 Action 会有不同的 Data 封装，例如打电话的 Intent 会封装 tel:格式的电话 URI，而 ACTION_VIEW 的 Intent 中的 Data 则会封装 http：格式的 URI。正确的 Data 封装对 Intent 匹配请求同样非常重要。

Category 是对目标组件类别信息的描述。同样作为一个字符串对象，一个 Intent 中可以包含多个 Category。Category 相关的方法有三个，addCategory()添加一个 Category，removeCategory()删除一个 Category，而 getCategory()得到一个 Category。Android 系统同样定义了一组静态字符常量来表示 Intent 的不同类型，下面列出一些常见的 Category 常量：

- CATEGORY_GADGET，表明目标 Activity 是可以嵌入到其他 Activity 中的；
- CATEGORY_HOME，表明目标 Activity 是 HOMEActivity；
- CATEGORY_TAB，表明目标 Activity 是 TabActivity 的一个标签下的 Activity；
- CATEGORY_LAUNCHER，表明目标 Activity 是应用程序中最先被执行的 Activity；
- CATEGORY_PREFERNCE，表明目标 Activity 是一个偏好设置的 Activity。

Extra 中封装了一些额外的附加信息，这些信息是以键值对的形式存在的。Intent 可以通过 putExtras()和 getExtras()方法来存储和获取 Extra。在 Android 系统 的 Intent 类中，同样对一些常用的 Extra 键值进行定义：

- EXTRA_BCC，装有邮件密送地址的字符串数组；
- EXTRA_EMAIL，装有邮件发送地址的字符串数组；
- EXTRA_UID，使用 ACTION_UID_REMOVED 动作时，描述删除用户的 id；
- EXTRA_TEXT，当使用 ACTION_SEND 动作时，描述要发送文本的信息。

Flag 是指一些有关系统如何启动组件的标志位，Android 同样对其进行了封装。

4.1.2 Activity 跳转及传值

在 Android 系统中，应用程序一般都有多个 Activity，Intent 可以实现不同 Activity 之间的切换和数据传递。Intent 启动 Activity 方式可以分为显式启动和隐式启动。显式启动必须在 Intent 中指明启动的 Activity 所在的类，而隐式启动则由 Android 系统根据 Intent 的动作和数据来决定启动哪一个 Activity，即在隐式启动时，Intent 中只包含需要执行的动作和所包含的数据，而无需指明具体启动哪一个 Activity，选择权由 Android 系统和最终用户来决定。

（1）界面跳转。使用 Intent 来显式启动 Activity，首先需要创造一个 Intent，并为它指定当前的应用程序上下文以及要启动的 Activity，把创建好的 Intent 作为参数传递给 startActivity()方法。具体格式如下：

```
// 实例化 Intent 对象,设置跳转 Activity
Intent intent = new Intent(MainActivity.this, SecondActivity.class);
// 进行跳转
startActivity(intent);
```

【实例 4.1】Intent 实现界面跳转的应用。

下面的程序简单说明如何使用 Intent 启动 activity。示例包含两个 activity 类，分别是 MainActivity.java 和 SecondActivity.java。程序 MainActivity 界面中有一个按钮，如图 4-1 所示，当单击按钮时跳转到 SecondActivity。程序默认启动 MainActivity 这个 Activity。

图 4-1　MainActivity 界面

步骤一：创建名为“IntentDemo”的工程，在 com.example.intentdemo 包下新建名为 SecondActivity 的 Activity 类。在创建 SecondActivity.java 类的同时，Eclipse 会在 res/layout 目录下为该 Activity 自动生成名为 activity_second.xml 的布局文件，并且在 AndroidManifest.xml 文件中注册该 Activity。

AndroidManifest.xml 文件代码如下：

```
<?xml version="1.0" encoding="utf-8"?>
<manifest xmlns:android="http://schemas.android.com/apk/res/android"
    package="com.example.intentdemo"
    android:versionCode="1"
    android:versionName="1.0" >

    <uses-sdk
        android:minSdkVersion="8"
        android:targetSdkVersion="17" />

    <application
        android:allowBackup="true"
        android:icon="@drawable/ic_launcher"
        android:label="@string/app_name"
        android:theme="@style/AppTheme" >
        <activity
            android:name="com.example.intentdemo.MainActivity"
            android:label="@string/app_name" >
            <intent-filter>
                <action android:name="android.intent.action.MAIN" />

                <category android:name="android.intent.category.LAUNCHER"/>
            </intent-filter>
        </activity>
        <activity
            android:name="com.example.intentdemo.SecondActivity"
            android:label="@string/title_activity_second" >
        </activity>
    </application>

</manifest>
```

步骤二：在 res/layout/activity_main.xml 中添加一个按钮，并通过 android:onClick 属性，为按钮指定监听事件，具体代码如下：

```
<RelativeLayout xmlns:android="http://schemas.android.com/apk/res/android"
    xmlns:tools="http://schemas.android.com/tools"
    android:layout_width="match_parent"
    android:layout_height="match_parent"
    android:paddingBottom="@dimen/activity_vertical_margin"
    android:paddingLeft="@dimen/activity_horizontal_margin"
    android:paddingRight="@dimen/activity_horizontal_margin"
    android:paddingTop="@dimen/activity_vertical_margin"
    tools:context=".MainActivity" >

    <Button
        android:layout_width="match_parent"
        android:layout_height="wrap_content"
        android:text="点我跳转到 SecondActivity 界面"
        android:onClick="myOnClick"/>

</RelativeLayout>
```

步骤三：在 MainActivity.java 中实现 activity_main.xml 布局中给 Button 指定 android:onClick="myOnClick"单击事件，并在 myOnClick 事件中使用 Intent 实现跳转，具体代码实现如下：

```
public class MainActivity extends Activity {

    @Override
    protected void onCreate(Bundle savedInstanceState) {
        super.onCreate(savedInstanceState);
        setContentView(R.layout.activity_main);
    }

    // 实现 myOnClick 单击事件
    public void myOnClick(View v) {
        // 实例化 Intent 对象,设置跳转 Activity
        Intent intent =
new Intent(MainActivity.this, SecondActivity.class);
        // 进行跳转
        startActivity(intent);
    }

}
```

在 myOnClick 单击事件的处理函数中，Intent 构造函数的第一个参数是应用程序上下文，程序中的应用程序上下文就是 IntentDemo；第二个参数是接收 Intent 的目标组件，使用的是显示启动方式，直接指明了需要启动的 Activity。

（2）界面传值。同样除了界面跳转也就是显示启动外，Intent 还可将当前界面的值传递到另一个界面。下面请看实例 4.2。

【实例 4.2】ntentValueDemo 界面传值应用。

本例将为大家简单介绍如何使用 Intent 进行传值。在这个例子中有两个 Activity，分别是 MainActivity 和 GetValueActivity，界面分别如图 4-2、图 4-3 所示。

图 4-2　MainActivity 界面

图 4-3　GetValueActivity

步骤一：创建名为“IntentValueDemo”的工程。打开 res/layout/activity_main.xml 布局文件，添加如图 4-2 所示的文本框和按钮，具体代码如下：

```
<RelativeLayout xmlns:android="http://schemas.android.com/apk/res/android"
    xmlns:tools="http://schemas.android.com/tools"
    android:layout_width="match_parent"
    android:layout_height="match_parent"
    android:paddingBottom="@dimen/activity_vertical_margin"
    android:paddingLeft="@dimen/activity_horizontal_margin"
    android:paddingRight="@dimen/activity_horizontal_margin"
    android:paddingTop="@dimen/activity_vertical_margin"
    tools:context=".MainActivity" >

    <EditText
        android:id="@+id/edt_value"
        android:layout_width="match_parent"
        android:layout_height="wrap_content" />

    <Button
        android:layout_width="match_parent"
        android:layout_height="wrap_content"
        android:layout_below="@+id/edt_value"
        android:onClick="myOnClick"
        android:text="点我跳转并传值" />

</RelativeLayout>
```

步骤二：创建名为 GetValueActivity 的界面，与此同时 Eclipse 会为该 activity 创建名为 activity_getvalue.xml 的布局文件，并且在 AndroidManifest.xml 中注册该 Activity。打开 res/layout/activity_getvalue.xml 文件，添加如图 4-3 所示的按钮和文本控件，具体代码如下：

```
<RelativeLayout xmlns:android="http://schemas.android.com/apk/res/android"
    xmlns:tools="http://schemas.android.com/tools"
    android:layout_width="match_parent"
    android:layout_height="match_parent"
    android:paddingBottom="@dimen/activity_vertical_margin"
```

```
    android:paddingLeft="@dimen/activity_horizontal_margin"
    android:paddingRight="@dimen/activity_horizontal_margin"
    android:paddingTop="@dimen/activity_vertical_margin"
    tools:context=".GetValueActivity" >

    <Button
        android:id="@+id/btn_getvalue"
        android:layout_width="match_parent"
        android:layout_height="wrap_content"
        android:onClick="myOnClick"
        android:text="点我获取传入的值" />

    <TextView
        android:id="@+id/txt_show"
        android:layout_width="match_parent"
        android:layout_height="wrap_content"
        android:layout_below="@+id/btn_getvalue"
        android:text="显示传入的值"
        android:gravity="center_horizontal"/>

</RelativeLayout>
```

步骤三：单击图 4-2 中的按钮，取得文本编辑框里面的值，放入 Intent 并传到 GetValueActivity 界面显示出来。此时打开 MainActivity.java 文件，实现在 res/layout/activity_main.xml 中给按钮指定的 myOnClick 单击事件，实现界面跳转的同时进行值的传递，具体代码如下：

```
public class MainActivity extends Activity {

    // 声明控件对象
    private EditText edtValue;

    @Override
    protected void onCreate(Bundle savedInstanceState) {
        super.onCreate(savedInstanceState);
        setContentView(R.layout.activity_main);
        // 获取控件资源
        edtValue = (EditText) findViewById(R.id.edt_value);
    }

    // 实现监听事件
    public void myOnClick(View v){
        // 获取用户输入的值
        String value = edtValue.getText().toString();
        // 实例化 Intent 对象,设置跳转 Activity
        Intent intent = new Intent(MainActivity.this, GetValueActivity.class);
        // 把值放入 Intent
```

```
            intent.putExtra("VALUE", value);
            // 开始跳转
            startActivity(intent);
        }
    }
```

步骤四：在 GetValueActivity 界面中，有一个按钮和一个文本框，当单击按钮时，通过 getIntent()方法取得从 MainActivity 界面传来的 Intnet 对象，并取得里面的值显示在文本框中。具体代码如下：

```
public class GetValueActivity extends Activity {

    // 声明控件对象
    private TextView txtShow;

    @Override
    protected void onCreate(Bundle savedInstanceState) {
        super.onCreate(savedInstanceState);
        setContentView(R.layout.activity_getvalue);
        // 获取控件资源
        txtShow = (TextView) findViewById(R.id.txt_show);
    }

    // 实现监听事件
    public void myOnClick(View v){
        // 获取 Intent 对象
        Intent intent = getIntent();
        // 取得 Intent 里面的值
        String value = intent.getStringExtra("VALUE");
        // 放入 textView
        txtShow.setText(value);
    }
}
```

4.1.3 调用其他程序中的 Activtiy

之前讲了 Intent 显式启动，下面给大家介绍隐式启动。其好处在于不需要指明要启动哪一个 Activity，而由 Android 系统来决定，这样有利于使用第三方组件。

隐式启动 Activity 时，Android 系统在应用程序运行时解析 Intent,并根据一定的规则对 Intent 和 Activity 进行配置，使 Intent 上的动作、数据与 Activity 完全吻合。匹配的 Activity 可以是应用程序本身的，也可以是 Android 系统内置的，还可以是第三方应用程序提供的。因此，这种方式更加强调了 Android 应用程序中组件的可复用性。

例如：用户希望启动一个浏览器，查看指定的网页内容，却不能确定具体应该启动哪一个 Activity 进行配置，此时可以使用 Intent 的隐式启动方式，由 Android 系统在程序运行时决定具体启动哪一个应用程序的 Activity 来接收这个 Intent。程序开发人员可以浏览动作和 Web 地址作为参数传给 Intent，Android 系统则通过匹配动作和数据格式，找到最适合于此动作的组件。在默认情况下，Android 系统会调用内置的 Web 浏览器。如下是访问百度网页的例子代码：

```
// 实例化 Intent 对象,设置跳转 Activity
Intent intent = new Intent(Intent.ACTION_VIEW,Uri.parse("http://www.baidu.com"));
// 进行跳转
startActivity(intent);
```

Intent 的动作是 Intent.ACTION_VIEW，根据 URI 的数据类型来匹配动作。数据部分的 URI 是 Web 地址，使用 Uri.parse(urlString)方法，可以简单地把一个字符串解释成 URI 对象。Intent 的语法如下：

```
Intent intent = new Intent(Intent.ACTION_VIEW, Uri.parse(urlString));
```

Intent 构造函数的第一个参数是 Intent 需要执行的动作，Android 系统支持的常见动作字符串常量可以参见表 4-1。第二个参数是 URI，表示需要传递的数据。

表 4-1 Intent 常用动作表

动作	说明
ACTION_ANSWER	打开接听电话的 Activity，默认为 Android 内置的拨号盘界面
ACTION_CALL	打开拨号盘界面并拨打电话，使用 URI 中的数字部分为电话号码
ACTION_DELETE	打开一个 Activity，对所提供的数据迚行删除操作
ACTION_DIAL	打开内置拨号盘界面，显示 URI 中提供的电话号码
ACTION_EDIT	打开一个 Activity，对提供的数据进行编辑操作
ACTION_INSERT	打开一个 Activity，在提供的数据的当前位置插入新项
ACTION_PICK	启动一个 Activity，从提供的数据列表中选取一项
ACTION_SEARCH	启动一个 Activity，执行搜索动作
ACTION_SENDTO	启动一个 Activity，向数据提供的联系人发送信息
ACTION_SEND	启动一个可以发送数据的 Activity
ACTION_VIEW	最常用的动作，对以 URI 方式传送的数据，根据 URI 协议部分以最佳方式启动相应的 Activity 进行处理。对于 http:address 将打开浏览器查看：对于 tel:address 将打开拨号呼叫指定的电话号码
ACTION_WEB_SERC H	打开一个 Activity，对提供的数据进行 Web 搜索

【实例 4.3】IntentToCallDemo 拨打电话应用。

本例将为大家演示如何调用系统拨号界面拨打电话。项目只有一个简单的界面，包含一个文本编辑框和一个按钮，界面如图 4-4 所示。

图 4-4　自定义拨号界面

步骤一：新建名为 IntentToCallDemo 的工程。打开 res/layout/activity_main.xml 文件，绘制图 4-4 所展示的界面，具体代码如下：

```
<LinearLayout xmlns:android="http://schemas.android.com/apk/res/android"
    xmlns:tools="http://schemas.android.com/tools"
    android:layout_width="match_parent"
    android:layout_height="match_parent"
    android:paddingBottom="@dimen/activity_vertical_margin"
    android:paddingLeft="@dimen/activity_horizontal_margin"
    android:paddingRight="@dimen/activity_horizontal_margin"
    android:paddingTop="@dimen/activity_vertical_margin"
    tools:context=".MainActivity" >

    <EditText
        android:id="@+id/edt_callnum"
        android:layout_width="wrap_content"
        android:layout_height="wrap_content"
        android:layout_weight="8"
        android:phoneNumber="true"
        android:hint="请输入电话号"/>

    <Button
        android:id="@+id/btn_call"
        android:layout_width="wrap_content"
        android:layout_height="wrap_content"
        android:layout_toRightOf="@+id/edt_callnum"
        android:text="Call"
        android:layout_weight="1"
        android:onClick="myOnClick"/>

</LinearLayout>
```

步骤二：当单击 Call 按钮时，获取输入的电话号，调用系统拨号界面完成拨打电话的功能。打开 MainActivity.java 文件，添加 Intent 隐式启动功能。具体代码如下：

```
public class MainActivity extends Activity {
    // 声明控件对象
    private EditText edtCallNum;

    @Override
    protected void onCreate(Bundle savedInstanceState) {
        super.onCreate(savedInstanceState);
```

```
        setContentView(R.layout.activity_main);
        // 获取控件资源
        edtCallNum = (EditText) findViewById(R.id.edt_callnum);
    }

    // 实现监听事件
    public void myOnClick(View v){
        // 获取用户输入的电话号
        String callNum = edtCallNum.getText().toString();
        // 实例化 Intent 对象，设置参数
        Intent intent = new Intent(Intent.ACTION_CALL, Uri.parse("tel:" + callNum));
        // 开始条状
        startActivity(intent);
    }
}
```

步骤三：最后需要在 AndroidManifest.xml 里面添加拨打电话的权限，具体代码如下：

```
<?xml version="1.0" encoding="utf-8"?>
<manifest xmlns:android="http://schemas.android.com/apk/res/android"
    package="com.example.intenttocalldemo"
    android:versionCode="1"
    android:versionName="1.0" >

    <uses-sdk
        android:minSdkVersion="8"
        android:targetSdkVersion="17" />
     <!-- 电话权限 -->
     <uses-permission android:name="android.permission.CALL_PHONE"/>

    <application
        android:allowBackup="true"
        android:icon="@drawable/ic_launcher"
        android:label="@string/app_name"
        android:theme="@style/AppTheme" >
        <activity
            android:name="com.example.intenttocalldemo.MainActivity"
            android:label="@string/app_name" >
            <intent-filter>
                <action android:name="android.intent.action.MAIN" />

                <category android:name="android.intent.category.LAUNCHER"/>
            </intent-filter>
        </activity>
    </application>

</manifest>
```

4.2 BroadcastReceiver 广播组件应用

Intent 对象不仅可以启动应用程序内部或其他应用程序的 Activity，它的另一种用途还可发送广播消息。应用程序和 Android 系统都可以使用 Intent 发送广播消息，广播消息的内容可以是和应用程序密切相关的数据信息，也可以是 Android 的系统信息，例如网络连接变化、电池电量变化、接收到短信和系统设置变化等等。实际上，在手机中发生这类事件时，Android 都会向整个系统发送相应的 Broadcast Action。如果应用程序注册了 BroadcastReceiver，则可以接收到指定的广播消息。BroadcastReceiver 类位于 android.content 包下，是对广播消息进行过滤并响应的控件。下面将详细介绍如何在应用程序中接收系统的 Broadcast 和向外发送广播。

4.2.1 接收广播消息

接收系统广播一般需要两步：

新建一个继承自 BroadcastReceiver 的类，并实现 BroadcastReceiver 类中的 onReceive 方法。如果应用程序接收到系统广播，就会调用 onReceive 方法。

在 AndroidManifest.xml 文件中使用<receiver>标签来指定在第一步中编写的接收系统广播的类可以接收哪一个 Broadcast Action，并同时添加系统权限。

完成上面的两步后，在手机或虚拟机上调试程序，然后退出程序。这时只要 Android 系统向外广播应用程序可以接收到的 Broadcast Action，并且程序未被卸载，系统会自动调用 onReceive 方法来处理这个 Broadcast Action。

【实例 4.4】BroadcastReceiverStartUp 接收系统广播消息应用。

程序示范了如何通过 BroadcastReceiver 接收系统广播达到每次开机都会启动的目的。该程序界面很简单，只有一个 TextView 用来显示一段文字，如图 4-5 所示。

图 4-5　开机启动首界面

步骤一：按照上述的内容新建类 StartUpReceiver 继承自 BroadcastReceiver 类，用于接收系统广播，具体代码如下：

```
public class StartUpReceiver extends BroadcastReceiver {

    @Override
    public void onReceive(Context context, Intent arg1) {
        // 实例化 intent 对象，设置参数
```

```
            Intent intent = new Intent(context, MainActivity.class);
            // 设置监听
            intent.setFlags(Intent.FLAG_ACTIVITY_NEW_TASK);
            // 开始跳转
            context.startActivity(intent);
    }

}
```

步骤二：同时在 AndroidManifest.xml 中配置 receiver 相关参数和权限，具体代码片段如下：

```
<?xml version="1.0" encoding="utf-8"?>
<manifest xmlns:android="http://schemas.android.com/apk/res/android"
    package="com.example.startup"
    android:versionCode="1"
    android:versionName="1.0" >

    <uses-sdk
        android:minSdkVersion="8"
        android:targetSdkVersion="17" />

    <application
        android:allowBackup="true"
        android:icon="@drawable/ic_launcher"
        android:label="@string/app_name"
        android:theme="@style/AppTheme" >
        <activity
            android:name="com.example.startup.MainActivity"
            android:label="@string/app_name" >
            <intent-filter>
                <action android:name="android.intent.action.MAIN" />

                <category android:name="android.intent.category.LAUNCHER"/>
            </intent-filter>
        </activity>

        <receiver android:name="StartUpReceiver">
            <intent-filter >
                <!-- 指定接收的 Broadcast Action -->
                <action android:name= "android.intent.action.BOOT_COMPLETED"/>
          <!-- 指定 Action 的种类。该种类表示 Android 系统启动后第一个运行的应用程序 -->
                <category android:name="android.intent.category.HOME"/>
            </intent-filter>
        </receiver>

    </application>
<!-- 注册权限 -->
<uses-permission android:name="android.permission.RECEIVE_BOOT_COMPLETED"/>
</manifest>
```

4.2.2 发送广播消息

使用 Intent 发送广播消息非常简单，只需创建一个 Intent，并调用 sendBroadcast()函数就可把 Intent 携带的消息广播出去。但需要注意的是，在构造 Intent 时必须用一个全局唯一的字符串标识其执行的动作，通常使用应用程序包的名称。

如果要在 Intent 传递额外数据，可以用 Intent 的 putExtra()方法。下面的代码构造用于广播消息的 Intent，并添加了额外的数据，然后调用 sendBroadcast()发出广播消息。

```
String ACTION_STRING = "com.example.sendmsg";
Intent intent = new Intent(ACTION_STRING);
intent.putExtra("key1", "value1");
intent.putExtra("key2", "value2");
sendBroadcast(intent);
```

【实例 4.5】BroadcastReceiverSendMsg 应用使用广播打开 BroadcastReceiverRecMsg 应用。

本例中简单示范了如何在 BroadcastReceiverSendMsg 程序中广播消息，并打开 BroadcastReceiverRecMsg 程序。BroadcastReceiverSendMsg 程序中只需一个简单的界面，包含一个按钮，如图 4-6 所示。

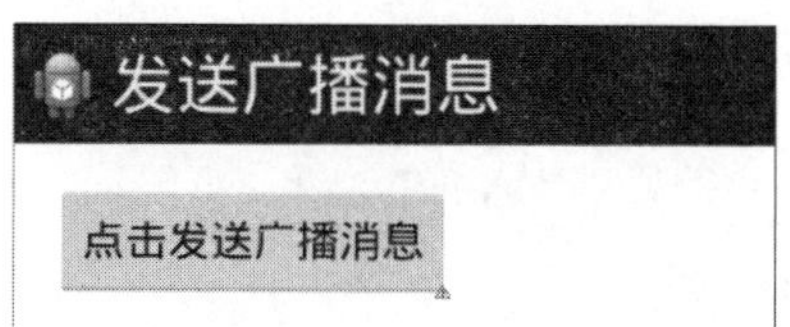

图 4-6 BroadcastReceiverSendMsg 界面

步骤一：创建名为 BroadcastReceiverSendMsg 的工程。打开 res/layout/activity_main.xml 文件，实现图 4-6 所展示的界面，具体代码如下：

```
<RelativeLayout xmlns:android="http://schemas.android.com/apk/res/android"
    xmlns:tools="http://schemas.android.com/tools"
    android:layout_width="match_parent"
    android:layout_height="match_parent"
    android:paddingBottom="@dimen/activity_vertical_margin"
    android:paddingLeft="@dimen/activity_horizontal_margin"
    android:paddingRight="@dimen/activity_horizontal_margin"
    android:paddingTop="@dimen/activity_vertical_margin"
    tools:context=".MainActivity" >

    <Button
        android:layout_width="wrap_content"
        android:layout_height="wrap_content"
        android:text="点击发送广播消息"
        android:onClick="myOnClick" />

</RelativeLayout>
```

步骤二： 当单击按钮时发出广播，启动 BroadcastReceiverRecMsg 程序。MainActivity.java 具体代码实现如下。

```
public class MainActivity extends Activity {
        @Override
        protected void onCreate(Bundle savedInstanceState) {
                super.onCreate(savedInstanceState);
                setContentView(R.layout.activity_main);
        }

        public void myOnClick(View v){
                Intent intent = new Intent("com.example.receivermsg.ACTION");
                sendBroadcast(intent);
        }
}
```

步骤三： 创建名为 BroadcastReceiverRecMsg 的工程，在该工程中只需仿照之前的开机启动程序，完成接收发出的广播即可，界面代码就不再展示。新建类 RecMsgReceiver 继承自 BroadcastReceiver 类，具体代码实现如下。

```
public class RecMsgReceiver extends BroadcastReceiver {
        // 声明常量定义 Broadcast ACTION
        private final String ACTION = "com.example.receivermsg.ACTION";

        @Override
        public void onReceive(Context context, Intent arg1) {
                // 判断传入的 ACTION
                if (arg1.getAction().equals(ACTION)) {
                        // 实例化 intent 对象，设置参数
                        Intent intent = new Intent(context, MainActivity.class);
                        // 设置监听
                        intent.setFlags(Intent.FLAG_ACTIVITY_NEW_TASK);
                        // 开始跳转
                        context.startActivity(intent);
                }
        }

}
```

步骤四： 在 AndroidManifest.xml 配置 receiver 相关参数，其他重复代码用省略号表示，具体实现代码请看加粗部分：

```
<manifest xmlns:android="http://schemas.android.com/apk/res/android"
    ………… >
    <application …………>
        <activity ………… >
            …………
        </activity>

        <receiver android:name="RecMsgReceiver" >
```

```
                <intent-filter>
                    <!-- 指定接收的 Broadcast Action -->
                  <action android:name="com.example.receivermsg.ACTION" />
                </intent-filter>
            </receiver>

        </application>
    </manifest>
```

完成以上内容，即可通过单击 BroadcastReceiverSendMsg 的按钮启动 BroadcastReceiverRecMsg 程序。

4.3 实训项目

使用 Intent 浏览网页

Android SDK 内置的 Web 浏览器也对外提供了 Action，可以通过调用这个 Action 来传递一个 Web 网址，并通过 Web 浏览器来打开这个 Web 网址。主界面只需一个文本编辑框和一个按钮，如图 4-7 所示。

图 4-7　浏览网页主界面

【实例 4.6】XXXXXXX 浏览网页应用。

步骤一：创建名为 XXXXXXX 的工程，实现图 4-7 的界面，代码不再展示。

步骤二：打开 MainActivity.java 文件，实现当单击按钮时获取用户输入的网址，并通过 Intent 打开浏览器，打开指定网页的功能。具体代码如下：

```
public class MainActivity extends Activity {
        // 声明控件对象
        private EditText edtWeb;
        @Override
        protected void onCreate(Bundle savedInstanceState) {
                super.onCreate(savedInstanceState);
                setContentView(R.layout.activity_main);
                // 获取控件资源
                edtWeb = (EditText) findViewById(R.id.edt_web);
        }
```

```

        // 实现监听事件
        public void myOnClick(View v){
            // 获取用户输入的网址
            String web = edtWeb.getText().toString();
            // 实例化 Intent 设置参数
            Intent webIntent = new Intent(Intent.ACTION_VIEW, Uri.parse(web));
            // 开始跳转
            startActivity(webIntent);
        }
    }
```

最终效果如图 4-8、图 4-9 所示。

图 4-8 输入网址的主界面

图 4-9 打开网页后的界面

4.4 本章小结

本章主要介绍了 Intent 对象在 Activity 和广播中的应用。大多数读者接触到的第一个关于 Intent 的应用就是利用 Intent 对象来启动 Activity。如果只想简单地启动 Activity，可以直接使用 startActivity 方法；如果想进行传值就需要用到 Intent 中的 putExtra 方法了，然后在跳转后的新界面中通过 getIntent 方法得到 Intent 里面的值。除此之外，还可使用 sendBroadcast 方法发送广播。这些广播实际上也是 Intent 对象，只是这些 Intent 对象指定的是 Broadcast Action，而不是 Activity Action。如果想接收系统广播或自己发送的广播，就需要继承 BroadcastReceiver 类。在该类的 onReceive 方法中可以获取到接收的广播数据，并进行相关处理。

4.5　本章习题

请完成以下作业，具体要求如下：

1．新建名为 IntentToWebStartUp 项目。

2．实现把应用安装到虚拟机或手机里后，重新启动手机，开机后自动运行 IntentToWebStartUp 应用打开“http://www.baidu.com”网站。

5

Android 数据存储

学习目标：

数据存储是应用程序最基本的问题，任何企业系统、应用软件都必须解决这一问题，数据存储必须以某种方式保存，不能丢失，并且能够有效、简便地使用和更新这些数据。通过本章的学习，我们将掌握 Android 中数据存储方式和数据共享。

【知识目标】

- 简单文本存储
- 文件保存
- SQLite 数据库
- 数据库访问及操作
- 数据共享

【技能目标】

- 掌握 SharedPreferences 的使用
- 掌握 File 文件保存的两种方式
- 掌握 SQLite 数据库的创建、表的创建及 CRUD 操作
- 掌握 SQLite 数据库适配器的创建和使用方法
- 熟悉 ContentProvider 数据共享的实现

所有应用程序都必然涉及数据的输入、输出，Android 应用也不例外，应用程序的参数设置、程序运行状态数据这些都需要保存到外部存储器上，这样系统关机之后数据才不会丢失。

Android 应用开发是使用 Java 语言来开发的，因此开发者在 Java IO 中的编程经验大部分都可“移植”到 Android 应用开发上，但 Android 系统还提供了一些专门的 IO API，通过这些 API 可以更有效地进行输入、输出。

如果应用程序只有少量数据需要保存，那么使用普通文件就可以了；但如果应用程序有大量数据需要存储、访问，就需要借助于数据库了，Android 系统内置了 SQLite 数据库，SQLite 数据库是一个真正轻量级的数据库，它没有后台进程，整个数据库就对应一个文件，这样可以非常方便地在不同设备之间移植。Andmid 不仅内置了 SQLite 数据库，而且为访问 SQLite 数据库提供了大量便捷的 API。本章将会详细介绍如何在 Android 应用中使用 SQLite 数据库。

5.1　数据存储一：SharedPreferences 简单存储

有些时候，应用程序有少量的数据需要保存，而且这些数据的格式很简单：都是普通的字符串、标量类型的值等，比如应用程序的各种配置信息（如是否打开音效、是否使用振动效果等）、小游戏的玩家积分（如扫雷英雄榜之类的）、用户的登录信息等，对于这种数据，Android 提供了 SharedPreferences 进行保存。

5.1.1　SharedPreferences 与 Editor 简介

SharedPreferences 保存的数据主要是类似于配置信息格式的数据，因此它保存的数据主要是简单类型的 key-value 键值对。

SharedPreferences 接口主要负责读取应用程序的 Preferences 数据，它提供了如下常用方法来访问 SharedPreferences 中的 key-value 对：

- boolean contains(String key)：判断 SharedPreferences 是否包含特定 key 的数据；
- abstract Map<String, ?> getAll()：获取 SharedPreferences 数据里全部的 key-value 对；
- boolean getXxx(String key, xxx defValue)：获取 SharedPreferences 数据里指定 key 对应的 value。如果该 key 不存在，返回默认值 defValue。其中 xxx 可以是 boolean、float、int、long、String 等各种基本数据类型的值。

SharedPreferences 接口本身并没有提供写入数据的能力，而是通过 SharedPreferences 的内部接口，SharedPreferences 调用 edit()方法即可获取它所对应的 Editor 对象。Editor 提供了如下方法来向 SharedPreferences 写入数据：

- SharedPreferences.Editor clear()：清空 SharedPreferences 里所有数据；
- SharedPreferences.Editor putXxx(String key, xxx value)：向 SharedPreferences 存入指定 key 对应的数据。其中 xxx 可以是 boolean、float、int、long、String 等各种基本类型的值；
- SharedPreferences.Editor remove(String key)：删除 SharedPreferences 里指定 key 对应的数据项；

- boolean commit()：当 Editor 编辑完成后，调用该方法提交修改。

SharedPreferences 本身是一个接口，程序无法直接创建 SharedPreferences 实例，只能通过 Context 提供的 getSharedPreferences(String name，int mode)方法来获取 SharedPreferences 实例，该方法的第二个参数支持如下几个值：

- Context.MODE_PRIVATE：指定该 SharedPreferences 数据只能被本应用程序读、写；
- Context.MODE_WORLD_READABLE：指定该 SharedPreferences 数据能被其他应用程序读，但不能写；
- Context.MODE_WORLD_WRITEABLE：指定该 SharedPreferences 数据能被其他应用程序读、写。

5.1.2 SharedPreferences 使用

【实例 5.1】SharedPreferenceDemo 实现数据读写应用。

下面的程序示范了如何向 SharedPreferences 中写入和读取数据，该程序的界面很简单，如图 5-1 所示。当用户第一次启动该应用程序时，输入完相关信息后，单击“保存”按钮保存用户信息。当退出程序再次运行程序时，会自动把之前保存的账号填写到对应的文本框里。

图 5-1　登录账号保存界面

步骤一：创建名为 SharedPreferencesDemo 的工程。

步骤二：打开 res/layout/activity_main.xml 布局文件，绘制如图 5-1 的界面。

步骤三：打开 MainActivity.java 文件，添加如下代码：

```
public class MainActivity extends Activity {

    // 声明控件对象
    private Button btnSave;
    private EditText edtUserName;

    @Override
    protected void onCreate(Bundle savedInstanceState) {
        super.onCreate(savedInstanceState);
        setContentView(R.layout.activity_main);
        // 获取控件资源
        btnSave = (Button) findViewById(R.id.btn_save);
        edtUserName = (EditText) findViewById(R.id.edt_username);
```

```
        // 给按钮绑定监听
        btnSave.setOnClickListener(new BtnLis());
        // 调用读取方法，取得用户名
        String userName = readInfo();
        // 把用户名放入
        edtUserName.setText(userName);
    }

    // 新建按钮监听事件
    private class BtnLis implements OnClickListener{
        @Override
        public void onClick(View arg0) {
            // 获取用户名
            String userName = edtUserName.getText().toString();
            // 调用保存方法，存入用户名
            writeInfo(userName);
        }
    }

    // 写入方法的实现
    private void writeInfo(String userName){
        // 创建 SharedPreferences 文件
        SharedPreferences preferences = getSharedPreferences("user", Context.MODE_PRIVATE);
        // 取得 SharedPreferences 编辑器
        Editor edit = preferences.edit();
        // 往编辑器里放入 key - value
        edit.putString("USERkeyNAME", userName);
        // 提交
        edit.commit();
    }

    // 读取方法的实现
    private String readInfo(){
        // 创建 SharedPreferences 文件
        SharedPreferences preferences = getSharedPreferences("user", Context.MODE_PRIVATE);
        // 获得文件里保存的字段内容
        String userName = preferences.getString("USERkeyNAME", "");
        // 返回值
        return userName;
    }

}
```

上面的程序中 readInfo()方法用于读取 SharedPreferences 数据，当程序所读取的 SharedPreferences 文件根本不存在时，程序也返回默认值，并不会抛出异常；程序中 writeInfo()方法用于写入 SharedPreferences 数据。

5.1.3 SharedPreferences 文件存储位置和格式

运行上面的程序，单击程序中“保存”按钮，程序将完成 SharedPreferences 写入，写入完成后打开 DDMS 时 File Explorer 面板，然后展开文件浏览树，看到如图 5-2 所示的窗口。

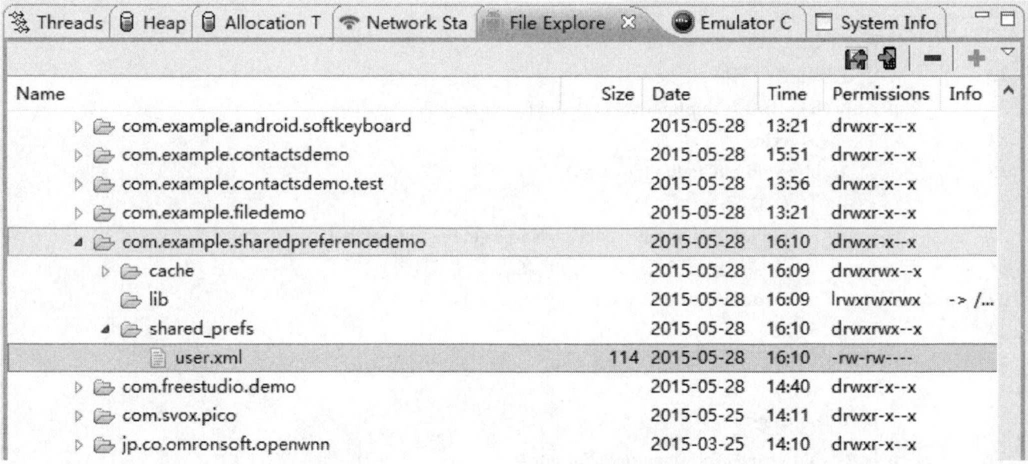

Name	Size	Date	Time	Permissions	Info
▷ com.example.android.softkeyboard		2015-05-28	13:21	drwxr-x--x	
▷ com.example.contactsdemo		2015-05-28	15:51	drwxr-x--x	
▷ com.example.contactsdemo.test		2015-05-28	13:56	drwxr-x--x	
▷ com.example.filedemo		2015-05-28	13:21	drwxr-x--x	
◢ com.example.sharedpreferencedemo		2015-05-28	16:10	drwxr-x--x	
▷ cache		2015-05-28	16:09	drwxrwx--x	
lib		2015-05-28	16:09	lrwxrwxrwx	-> /...
◢ shared_prefs		2015-05-28	16:10	drwxrwx--x	
user.xml	114	2015-05-28	16:10	-rw-rw----	
▷ com.freestudio.demo		2015-05-28	14:40	drwxr-x--x	
▷ com.svox.pico		2015-05-25	14:11	drwxr-x--x	
▷ jp.co.omronsoft.openwnn		2015-03-25	14:10	drwxr-x--x	

图 5-2 SharedPreferences 文件存储位置

SharedPreferences 数据文件保存在 /data/data/<package nam>/ shared_prefs 目录下，SharedPreferences 数据总是以 XML 格式保存。通过 File Explorer 面板的导出文件按钮将该 XML 文件导出到桌面，打开该 XML 文档可看到如下文件内容：

```
<?xml version='1.0' encoding='utf-8' standalone='yes' ?>
<map>
<string name="USERkeyNAME">jiandan</string>
</map>
```

从上面的文件不难看出，SharedPreferences 数据文件是一个根元素为<map.../>的根元素，该元素里每个子元素代表一个 key-value 对，当 value 是整数类型的值时使用<int.../>子元素，当 value 是字符串类型时，使用<string.../>子元素……依此类推。

5.2 数据存储二：File 文件存储

在 Java 中提供了一套完整的 IO 流体系，包括 FileInputStream、FileOutputStream 等，通过这些 IO 流可以非常方便地访问磁盘上的文件内容。Android 也支持以这种方式来访问手机存储器上的文件。

5.2.1 文件保存到 ROM

Context 提供了如下两个方法来打开应用程序的数据文件夹里的文件 IO 流：

- FileInputStream openFileInput(String name)：打开应用程序的数据文件夹下的 name 文件对应输入流；
- FileOutputStream openFileOutput(String name, int mode)：打开应用程序的数据文件夹下的 name 文件对应输出流。

上面两个方法分别用于打开文件输入流、输出流，其中第二个方法的第二个参数指定打开文件的模式，该模式支持如下值：

MODE.PRIVATE：该文件只能被当前程序读写；

MODE_APPEND：以追加方式打开该文件，应用程序可以向该文件中追加内容；

MODE_WORLD_READABLE：该文件的内容可以被其他程序读取；

MODE_WORLD_WRITEABLE：该文件的内容可由其他程序读、写。

除此之外，Context 还提供了如下两个方法来访问应用程序的数据文件夹：

getDir(String name, int mode)：在应用程序的数据文件夹下获取或创建 name 对应的子目录；

File getFilesDir()：获取该应用程序的数据文件夹的绝对路径；

String[] fileList()：返回该应用程序的数据文件夹下的全部文件；

deleteFile(String)：删除该应用程序的数据文件夹下的指定文件。

5.2.2　openFileOutput 和 openFileInput 使用

【实例 5.2】FileToROM 文件保存到 ROM 应用。

下面的程序简单示范了如何读写应用程序文件夹内的数据文件。该程序的界面布局同样很简单，只包含两个文本编辑框、两个按钮和一个文本框；其中第一组文本编辑框和按钮用于处理写入，文本编辑框用于接受用户输入，当用户按下“写入”按钮时，程序将会把数据写入文件；第二组文本框和按钮处理读取，文本框用于显示读取到的文件数据，当用户按下“读取”按钮时，该文本框显示文件中的数据。界面如图 5-3 所示。

图 5-3　文件存储界面

步骤一：创建名为 FileToROM 的工程。

步骤二：打开 res/layout/activity_main.xml 文件，绘制如图 5-3 所示的界面。具体代码如下：

```
<LinearLayout xmlns:android="http://schemas.android.com/apk/res/android"
    android:layout_width="match_parent"
    android:layout_height="match_parent"
    android:orientation="vertical" >

    <EditText
        android:id="@+id/edtname"
        android:layout_width="match_parent"
        android:layout_height="wrap_content"
        android:layout_marginTop="10dp"
        android:hint="请输入文件名" />

    <EditText
        android:id="@+id/edtcontent"
        android:layout_width="match_parent"
        android:layout_height="wrap_content"
        android:hint="请输入文件内容"
        android:minLines="5"/>

    <Button
        android:id="@+id/btnsave"
        android:layout_width="match_parent"
        android:layout_height="wrap_content"
        android:onClick="myOnClick"
        android:text="写入" />

    <TextView
        android:id="@+id/txtshow"
        android:layout_width="match_parent"
        android:layout_height="wrap_content"
        android:layout_marginLeft="4dp"
        android:layout_marginRight="4dp"
        android:background="#000000"
        android:hint="这里显示读取的文件内容"
        android:minLines="5"
        android:textColor="#ffffff"
        android:layout_marginTop="10dp"/>

    <Button
        android:id="@+id/btnread"
        android:layout_width="match_parent"
        android:layout_height="wrap_content"
        android:onClick="myOnClick"
        android:text="读取" />

</LinearLayout>
```

步骤三：打开 MainActivity.java 文件，在程序中添加如下代码：

```
public class MainActivity extends Activity {

        // 声明控件
        EditText edtFileName, edtFileContent;
        TextView txtShow;

        @Override
        protected void onCreate(Bundle savedInstanceState) {
                super.onCreate(savedInstanceState);
                setContentView(R.layout.activity_main);
                // 获取控件资源
                edtFileName = (EditText) findViewById(R.id.edtname);
                edtFileContent = (EditText) findViewById(R.id.edtcontent);
                txtShow = (TextView) findViewById(R.id.txtshow);
        }

        // 实现按钮监听事件
        public void myOnClick(View v) throws IOException {
                // 获取文件名
                String name = edtFileName.getText().toString();
                // 判断按钮
                switch (v.getId()) {
                // 保存按钮功能
                case R.id.btnsave:
                        // 获取文件内容
                        String content = edtFileContent.getText().toString();
                        // 判断文件名和内容是否为空，如果为空，则给出提醒
                        if (name.length() == 0 || content.length() == 0) {
                                Toast.makeText(this, "名字或内容不能为空!", Toast.LENGTH_SHORT).show();
                        } else {
                                try {
                                        // 调用保存文件的方法
                                        saveToFile(name, content);
                                        Toast.makeText(this, "保存成功！", Toast.LENGTH_SHORT).show();
                                } catch (IOException e) {
                                        e.printStackTrace();
                                        Toast.makeText(this, "保存失败！", Toast.LENGTH_SHORT).show();
                                }
                        }
                        break;
                // 读取按钮功能
                case R.id.btnread:
                        // 调用读取文件的方法
                        String readCon = readToFile(name);
                        // 显示在文本框中
```

```
                    txtShow.setText(readCon);
                    break;
                default:
                    break;
            }
        }

        // 保存文件的方法
        private void saveToFile(String name, String content) throws IOException {
            // 实例化 FileOutputStreanm
            FileOutputStream fos = openFileOutput(name, Context.MODE_APPEND);
            // 写入文件
            fos.write(content.getBytes());
            // 关闭流
            fos.close();
        }

        // 读取文件的方法
        private String readToFile(String name) throws IOException {
            // 实例化 FileInputStream
            FileInputStream fis = openFileInput(name);
            // 使用字节流转换数据，保存读取的数据
            ByteArrayOutputStream baos = new ByteArrayOutputStream();
            byte[] buffer = new byte[1024];
            int len;
            while((len = fis.read(buffer)) != -1)
                baos.write(buffer, 0, len);
            // 关闭流
            fis.close();
            baos.close();
            return baos.toString();
        }

}
```

上面的程序中 readToFile(String name)方法用于读取应用程序的数据文件，saveToFile(String name, String content)用于向应用程序中写入数据文件。从上面的代码可以看出，当 Android 系统调用 Context 的 openFileInput()、openFileOutput()打开文件输入流或输出流之后，接下来 IO 流的用法与 Java SE 中 IO 流的用法完全一样：想直接用节点读写也行，用包装流包装之后再处理也没问题。

5.2.3 ROM 文件存储位置

当按下程序中的“写入”按钮时，用户在文本框中输入的内容将会被保存到应用程序的数据文件中，打开 File Explorer 面板后，可以看到如图 5-4 所示的界面。

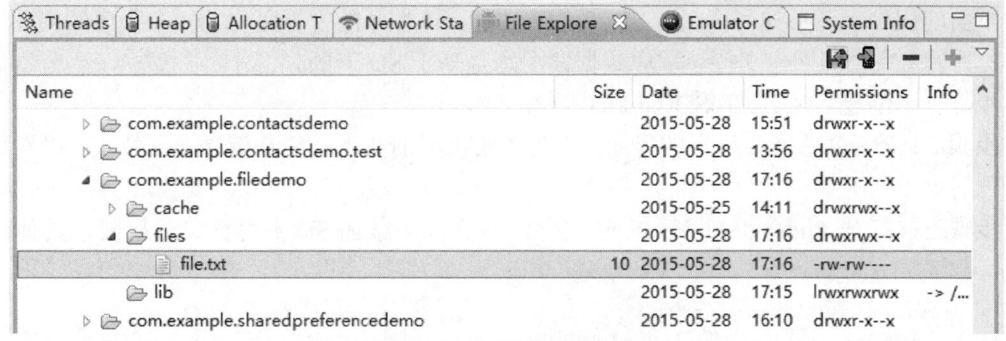

图 5-4　File 文件存储位置

应用程序的数据文件默认保存在/data/data/<packagename>/files 目录下。我们可以用导出 shared 文件的方法，把保存好的文件导出到桌面进行查看。

5.2.4　文件保存到 SDCard

当程序通过 Context 的 openFileInput 或 openFileOutput 来打开文件输入流、输出流时，程序所打开的都是应用程序的数据文件夹里的文件，这样所存储的文件大小可能比较有限，毕竟手机内置的存储空间是有限的。

为了更好地存、取应用程序的大文件数据，应用程序需要读、写 SD 卡上的文件。SD 卡大大扩充手机的存储能力。

读、写 SD 上的文件请按如下步骤进行：

（1）调用 Environment 的 getExternalStorageState()方法判断手机上是否插入 SD 卡，并且应用程序具有读写 SD 卡的权限。例如使用如下代码：

Environment.getExternalStorageState().equals(Environment.MEDIA_MOUNTED)

如果上面语句返回 true，表示手机已插入 SD 卡，且应用程序具有读写 SD 卡的能力。

（2）调用 Environment 的 getExternalStorageDirectory()方法来获取外部存储器根目录，也就是 SD 卡的目录。

（3）使用 FileInputStream、FileOutputStream 来读、写 SD 卡里的文件，并在 AndroidManifest.xml 添加外存储设备的读写权限。

应用程序读、写 SD 卡的文件有如下两个注意点：

1）手机上应该已插入 SD 卡。对于模拟器来说，可通过 mksdcard 命令来创建虚拟存储卡，或者在创建虚拟机时为虚拟机分配 SDCard 存储空间。

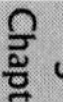

2）为了读、写 SD 卡上的数据，必须在应用程序的清单文件（AndroidManifest.xml）中添加读、写 SD 卡的权限。具体配置如下：

```
<!-- 向 SDCard 写入数据的权限 -->
<uses-permission android:name="android.permission.WRITE_EXTERNAL_STORAGE" />
```

【实例 5.3】FileToSDCard 文件保存到 SDCard 应用。

下面的程序示范了如何读、写 SD 卡上的文件，该程序的主界面与上一个程序的界面（如图 5-3 所示）完全相同。支持该程序数据读、写是基于 SD 卡的。

步骤一：创建名为 FileToSDCard 的工程。

步骤二：绘制如图 5-3 所示的界面，因为本实例界面与上一个实例界面一致，所以布局代码不再展示。

步骤三：打开 AndroidManifest.xml 文件，在文件中添加 SD 卡可读、写权限，具体代码如下：

```
<?xml version="1.0" encoding="utf-8"?>
<manifest xmlns:android="http://schemas.android.com/apk/res/android"
    package="com.example.filedemo"
    android:versionCode="1"
    android:versionName="1.0" >

    <uses-sdk
        android:minSdkVersion="8"
        android:targetSdkVersion="17" />

      <!-- 向 SDCard 写入数据的权限 -->
      <uses-permission android:name="android.permission.WRITE_EXTERNAL_STORAGE" />

    <instrumentation
        android:name="android.test.InstrumentationTestRunner"
        android:targetPackage="com.example.filedemo" />

    <application
        android:allowBackup="true"
        android:icon="@drawable/ic_launcher"
        android:label="@string/app_name"
        android:theme="@style/AppTheme" >
        <uses-library android:name="android.test.runner" />

        <activity
            android:name="com.example.filedemo.MainActivity"
            android:label="@string/app_name" >
            <intent-filter>
                <action android:name="android.intent.action.MAIN" />

                <category android:name="android.intent.category.LAUNCHER" />
            </intent-filter>
        </activity>
    </application>

</manifest>
```

步骤四：打开 MainActivity.java 文件，添加如下代码：

```
public class MainActivity extends Activity {

    // 声明控件
    EditText edtFileName, edtFileContent;
    TextView txtShow;

    @Override
    protected void onCreate(Bundle savedInstanceState) {
        super.onCreate(savedInstanceState);
        setContentView(R.layout.activity_main);
        // 获取控件资源
        edtFileName = (EditText) findViewById(R.id.edtname);
        edtFileContent = (EditText) findViewById(R.id.edtcontent);
        txtShow = (TextView) findViewById(R.id.txtshow);
    }

    // 实现按钮监听事件
    public void myOnClick(View v) throws IOException {
        // 获取文件名
        String name = edtFileName.getText().toString();
        // 判断按钮
        switch (v.getId()) {
        // 保存按钮功能
        case R.id.btnsave:
            // 获取文件内容
            String content = edtFileContent.getText().toString();
            // 判断文件名和内容是否为空，如果为空，则给出提醒
            if (name.length() == 0 || content.length() == 0) {
                Toast.makeText(this, "名字或内容不能为空!", Toast.LENGTH_SHORT).show();
            } else {
                // 判断是否插入 SDCard
                if(Environment.getExternalStorageState().equals(Environment.MEDIA_MOUNTED)){
                    try {
                        // 调用保存文件的方法
                        saveToSDCard(name, content);
                        Toast.makeText(this, "保存成功！", Toast.LENGTH_SHORT).show();
                    } catch (IOException e) {
                        e.printStackTrace();
                        Toast.makeText(this, "保存失败！", Toast.LENGTH_SHORT).show();
                    }
                } else {
                    Toast.makeText(this, "SDCard 异常！", Toast.LENGTH_SHORT).show();
                }
            }
            break;
        // 读取按钮功能
```

```
                case R.id.btnread:
                    // 调用读取文件的方法
                    String readCon = readToSDCard(name);
                    // 显示在文本框中
                    txtShow.setText(readCon);
                    break;
                }
        }

        // 保存文件的方法
        private void saveToSDCard(String name, String content) throws IOException
        {
                // 获取 SD 卡所在目录(兼容所有版本)
                File file = new File(Environment.getExternalStorageDirectory(), name);
                // 实例化 FileOutputStreanm，创建输出流，指向 SD 卡所在目录
                FileOutputStream fos = new FileOutputStream(file);
                // 写入文件
                fos.write(content.getBytes());
                // 关闭流
                fos.close();
        }

        // 读取文件的方法
        private String readToSDCard(String name) throws IOException {
                // 获取 SD 卡所在目录(兼容所有版本)
                File file = new File(Environment.getExternalStorageDirectory(), name);
                // 实例化 FileInputStream,创建输入流，指向 SD 卡所在目录
                FileInputStream fis = new FileInputStream(file);
                // 使用字节流转换数据，保存读取的数据
                ByteArrayOutputStream baos = new ByteArrayOutputStream();
                byte[] buffer = new byte[1024];
                int len;
                while((len = fis.read(buffer)) != -1)
                        baos.write(buffer, 0, len);
                // 关闭流
                fis.close();
                baos.close();
                return baos.toString();
        }

}
```

上面的程序中 readToSDCard(String name)用于读取 SD 卡中指定文件的内容；saveToSDCard(String name, String content)用于向 SD 卡写入文件。运行上面的程序，在第一个文本框内输入一些文本，然后单击“写入”按钮即可将数据写入底层 SD 卡。单击“读取”按钮又可把之前保存到文件的内容读取到文本框中，并显示出来。

5.2.5　SDCard 文件存储位置

打开 File Explorer，即可看到如图 5-5 所示的界面。

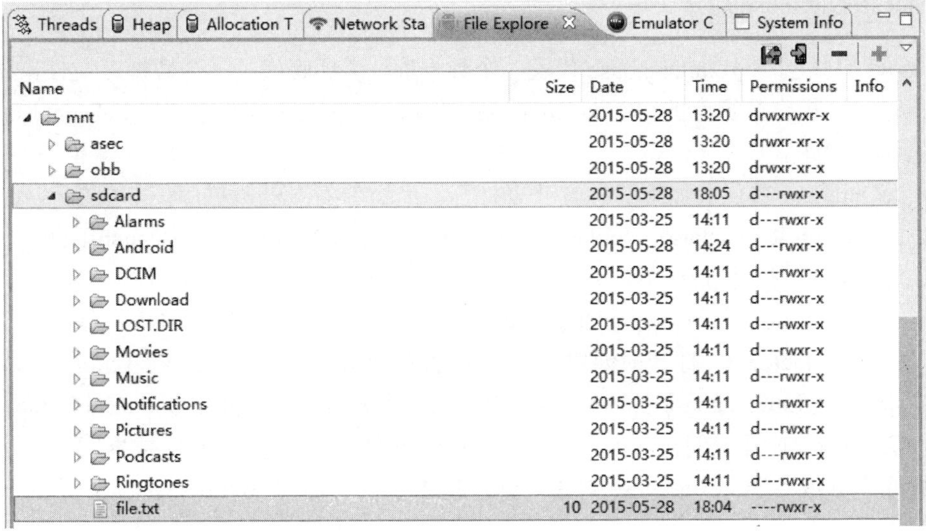

Name	Size	Date	Time	Permissions	Info
mnt		2015-05-28	13:20	drwxrwxr-x	
asec		2015-05-28	13:20	drwxr-xr-x	
obb		2015-05-28	13:20	drwxr-xr-x	
sdcard		2015-05-28	18:05	d---rwxr-x	
Alarms		2015-03-25	14:11	d---rwxr-x	
Android		2015-05-28	14:24	d---rwxr-x	
DCIM		2015-03-25	14:11	d---rwxr-x	
Download		2015-03-25	14:11	d---rwxr-x	
LOST.DIR		2015-03-25	14:11	d---rwxr-x	
Movies		2015-03-25	14:11	d---rwxr-x	
Music		2015-03-25	14:11	d---rwxr-x	
Notifications		2015-03-25	14:11	d---rwxr-x	
Pictures		2015-03-25	14:11	d---rwxr-x	
Podcasts		2015-03-25	14:11	d---rwxr-x	
Ringtones		2015-03-25	14:11	d---rwxr-x	
file.txt	10	2015-05-28	18:04	----rwxr-x	

图 5-5　存入 SD 卡的数据文件

如果开发者不想使用 Environment.getExtemalStorageDirectory()这么复杂的语句来获取 SD 卡的路径，完全可以使用/mnt/sdcard/路径下代表 SD 卡的路径，然后程序通过判断/mnt/sdcard/路径是否存在就可知道手机是否已插入了 SD 卡。当然不推荐这样做，因为 android 有很多版本，各个版本可能对于 SDCard 路径都有不同的定义，所以建议还是用兼容性强点的 Environment.getExternalStorageDirectory()语句来获取 SDCard 路径。

5.3　数据存储三：SQLite 数据库

Android 系统集成了一个轻量级的数据库：SQLite。SQLite 并不想成为像 Oracle、MySQL 那样的专业数据库，它只是一个嵌入式的数据库引擎，适用于资源有限的设备（如手机、PDA 等）进行适量的数据存取。

虽然 SQLite 支持绝大部分 SQL 92 语法，也允许开发者使用 SQL 语句操作数据库中的数据，但 SQLite 并不像 Oracle、MySQL 数据库那样需要安装、启动服务器进程，SQLite 数据库只是一个文件。

从本质上来看，SQLite 的操作方式只是一种更为便捷的文件操作。后面我们会看到，当应用程序创建或打开一个 SQLite 数据库时，其实只是打开一个文件准备读写，因此有人说 SQLite 有点像 Microsoft 的 Access，但实际上 SQLite 功能要强大得多。可能有读者会问，如果实际项目中有大量数据需要读写，而且需要面临大量用户的并发存储怎么办呢？这种情况下，不应该把数据存放在手机的 SQLite 数据库里，毕竟手机还是手机，它的存储能力、计算能力都不足以让它充当服务器的角色。

5.3.1 SQLiteDatabase 简介

Android 提供了 SQLiteDatabase 代表一个数据库，说到底就是一个数据库文件，一旦应用程序获得了代表指定数据库的 SQLiteDatabase 对象，接下来就可通过 SQLiteDatabase 对象来管理、操作数据库了。

SQLiteDatabase 提供了如下静态方法来打开一个文件对应的数据库：

- static SQLiteDatabase openDatabase(String path, SQLiteDatabase.CursorFactory factory, int flags)：打开 path 文件所代表的 SQLite 数据库；
- static SQLiteDatabase openOrCreateDatabase(File file, SQLiteDatabase.Curso rFactoryfactory)：打开或创建（如果不存在）file 文件所代表的 SQLite 数据库；
- static SQLiteDatabase openOrCreateDatabase(String path,SQLiteDatabase.Cursor Factory factory)：打开或创建（如果不存在）path 文件所代表的 SQLite 数据库。

在程序中获取 SQLiteDatabase 对象之后，接下来就可调用 SQLiteDatabase 的如下方法来操作数据库了：

- execSQL(String sql, Object[] bindArgs)：执行带占位符的 SQL 语句；
- execSQL(String sql)：执行 SQL 语句；
- insert(String table, String nullColumnHack, ContentValues values)：执行表中插入数据；
- update(String table, ContentValues values, String whereClause, String[] whereArgs)：更新指定表中的特定数据；
- delete(String table, String whereClause, String[] whereArgs)：删除指定表中的特定数据；
- Cursor query(String table, String[] columns, String selection, String[]selection Args, String groupBy, String having, String orderBy)：对执行数据表执行查询；
- Cursor query(String table, String[] columns, String selection, String[] selectionArgs, String groupBy, String having, String orderBy, String limit)：对执行数据表执行查询。limit 参数控制最多查询几条记录（用于控制分页的参数）；
- Cursor query(boolean distinct, String table, String[] columns, String selection, String[] selectionArgs, String groupBy, String having, String orderBy, String limit)：对指定表执行查询语句。其中第一个参数控制是否去除重复值；
- rawQuery(String sql, String[] selectionArgs)：执行带占位符的 SQL 查询；
- beginTransaction()：开始事务；
- endTransaction()：结束事务。

从上面的方法不难看出，其实 SQLiteDatabase 的作用有点类似于 JDBC 的 Connection 接口，但 SQLiteDatabase 提供的方法更多：比如 insert、update、delete、query 等方法，其实这些方法完全可通过执行 SQL 语句来完成，但 Android 考虑到部分开发者对 SQL 语法不熟悉，所以提供这些方法帮助开发者以更简单的方式来操作数据表的数据。

上面查询方法都是返回一个 Cursor 对象，Android 中的 Cursor 类似于 JDBC 的 ResultSet，Cursor 同样提供了如下方法来移动查询结果的记录指针：

- move(int offset)：将记录指针向上或向下移动指定的行数。offset 为正数就是向下移动；为负数就是向上移动。
- boolean moveToFirst()：将记录指针移动到第一行，如果移动成功则返回 true；
- boolean moveToLast()：将记录指针移动到最后一行，如果移动成功则返回 true；
- boolean moveToNext()：将记录指针移动到下一行，如果移动成功则返回 true；
- boolean moveToPosition(int position)：将记录指针移动到指定的行，如果移动成功则返回 true；
- boolean moveToPrevious()：将记录指针移动到上一行，如果移动成功则返回 true。

一旦将记录指针移动到指定行之后，接下来就可以调用 Cursor 的 getXXX()方法获取该行的指定列的数据。

5.3.2 创建数据库和表

上面介绍了关于 SQLiteDatabase 类及相关方法的使用，除此之外 Android 平台还提供了一个数据库辅助类来创建或打开数据库，这个辅助类是 SQLiteOpenHelper。在该类的构造器中，调用 Context 中的方法创建并打开一个指定名称的数据库对象。继承和扩展 SQLiteOpenHelper 类主要做的工作就是重写以下两个方法：

- onCreate(SQLiteDatabase db)：当数据库被首次创建时执行该方法，一般将创建表等初始化操作在该方法中执行；
- onUpgrade(SQLiteDatabse dv, int oldVersion,int new Version)：当打开数据库时传入的版本号与当前的版本号不同时会调用该方法。

除了上述两个必须要实现的方法外，还可以选择性地实现 onOpen 方法，该方法会在每次打开数据库时被调用。

SQLiteOpenHelper 类的基本用法是：当需要创建或打开一个数据库并获得数据库对象时，首先根据指定的文件名创建一个辅助对象，然后调用该对象的 getWritableDatabase 或 getReadableDatabase 方法获得 SQLiteDatabase 对象。

调用 getReadableDatabase 方法返回的并不总是只读数据库对象，一般来说该方法和 getWriteableDatabase 方法的返回情况相同，只有在数据库仅开放只读权限或磁盘已满时才会返回一个只读的数据库对象。

下面通过一个简单的实例说明如何通过 SQLiteOpenHelper 类来创建数据库。

【实例 5.4】SQLiteDemo 创建数据库应用。

步骤一： 创建名为 SQLiteDemo 的工程。

步骤二： 在项目中新建名为 DBOpenHelper 的一个类，继承自 SQLiteOpenHelper，同时会在该类中实现创建和修改数据库的方法，具体代码如下：

```
//定义工具类, 继承 SQLiteOpenHelper
public class DBOpenHelper extends SQLiteOpenHelper {

        // 创建对象的时候, 需要传入上下文环境
        public DBOpenHelper(Context context) {
                super(context, "user.db", null, 2);
                /*
                 * 由于父类没有无参构造函数, 必须显式调用有参的构造函数
                 * 参数 1: 上下文环境, 用来确定数据库文件存储的目录
                 * 参数 2: 数据库文件的名字
                 * 参数 3: 生成游标的工厂, 填 null 就是使用默认的
                 * 参数 4: 数据库的版本, 从 1 开始
                 */
        }

        @Override
        public void onCreate(SQLiteDatabase db) {
                System.out.println("onCreate");
                // 执行 SQL 语句, 创建表
                db.execSQL("CREATE TABLE user
                                (id INTEGER PRIMARY KEY AUTOINCREMENT, username)");
        }

        @Override
        public void onUpgrade(SQLiteDatabase db, int oldVersion, int newVersion) {
                System.out.println("onUpgrade");
                // 执行 SQL 语句, 添加表
                db.execSQL("ALTER TABLE user ADD password INTEGER");
        }

}
```

上面的代码就用于打开或创建一个 SQLite 数据库，其中通过设置构造方法里的四个参数创建数据库，onCreate()方法用来创建数据库中表，onUpgrade()方法用来修改数据库中的表。当第一次调用 DBOpenHelper 类时，同时会创建数据库，并执行 onCreate()方法创建数据库表。如果需要修改表时，以 1 为单位增加构造方法里的第四个参数，再次调用 DBOpenHelper 类时，会执行 onUpgrade()方法修改数据库表。

步骤三： 打开 MainActivity.java 文件，在 onCreate()方法里添加调用 DBOpenHelper 的方法，创建并获取数据库实例，具体代码如下：

```
public class MainActivity extends Activity {

        @Override
```

```
        protected void onCreate(Bundle savedInstanceState) {
                super.onCreate(savedInstanceState);
                setContentView(R.layout.activity_main);
                // 实例化 DBOpenHelper 对象，创建数据库
                DBOpenHelper helper = new DBOpenHelper(this);
                // 获取数据库可写对象
                SQLiteDatabase db = helper.getWritableDatabase();
        }

}
```

步骤四：把应用运行到 AVD 中后，在 DDMS 中打开 File Explorer 导出数据库来查看，具体详见 5.3.4 节。

5.3.3 使用 SQL 语句操作 SQLite 数据库

SQLiteDatabase 中的 execSQL 方法可执行任意 SQL 语句，包括带占位符的 SQL 语句。但由于该方法没有返回值，一般用于执行 DDL 语句或 DML 语句；如果需要执行查询语句，则可调用 SQLiteDatabase 的 rawQuery(String sql, String[] selectionArgs)方法。例如，如下代码可用于执行 DML 语句：

```
//执行插入 SQL 语句
db.execSQL("INSERT INTO user(name, password) VALUES(?, ?)", new Object[] { "admin", "123456" });
```

【实例 5.5】SQLiteCRUDDemo 数据库的增删改查应用。

下面的程序示范了如何在 Android 应用中操作 SQLite 数据库，该程序提供了两个文本编辑框、四个按钮和文本框，用户可以在这两个文本编辑框中输入内容，当用户单击“插入”“删除”“修改”和“查询”按钮时查看其功能，如图 5-6 所示。

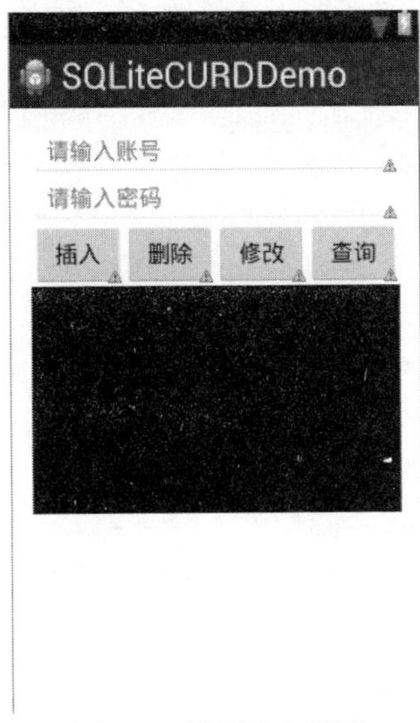

图 5-6 增删改查界面

步骤一：创建名为 SQLiteCRUDDemo 的工程。

步骤二：打开 res/layout/activity_main.xml 文件，搭建上述的界面，具体代码如下：

```
<LinearLayout xmlns:android="http://schemas.android.com/apk/res/android"
    android:layout_width="match_parent"
    android:layout_height="match_parent"
    android:orientation="vertical"
    android:paddingBottom="@dimen/activity_vertical_margin"
    android:paddingLeft="@dimen/activity_horizontal_margin"
    android:paddingRight="@dimen/activity_horizontal_margin"
    android:paddingTop="@dimen/activity_vertical_margin" >

    <EditText
        android:id="@+id/edtusername"
        android:layout_width="match_parent"
        android:layout_height="wrap_content"
        android:hint="请输入账号" />

    <EditText
        android:id="@+id/edtpassword"
        android:layout_width="match_parent"
        android:layout_height="wrap_content"
        android:hint="请输入密码" />

    <LinearLayout
        android:layout_width="match_parent"
        android:layout_height="wrap_content"
        android:orientation="horizontal" >

        <Button
            android:id="@+id/btninsert"
            android:layout_width="wrap_content"
            android:layout_height="wrap_content"
            android:layout_weight="1"
            android:onClick="myOnClick"
            android:text="插入" />

        <Button
            android:id="@+id/btndel"
            android:layout_width="wrap_content"
            android:layout_height="wrap_content"
            android:layout_weight="1"
            android:onClick="myOnClick"
            android:text="删除" />

        <Button
            android:id="@+id/btnupdate"
            android:layout_width="wrap_content"
```

```
                android:layout_height="wrap_content"
                android:layout_weight="1"
                android:onClick="myOnClick"
                android:text="修改" />

            <Button
                android:id="@+id/btnquery"
                android:layout_width="wrap_content"
                android:layout_height="wrap_content"
                android:layout_weight="1"
                android:onClick="myOnClick"
                android:text="查询" />
        </LinearLayout>

        <TextView
            android:id="@+id/txtshow"
            android:layout_width="match_parent"
            android:layout_height="wrap_content"
            android:background="#000000"
            android:minLines="10"
            android:textColor="#ffffff" />

</LinearLayout>
```

步骤三：构建用户实体类 User.java，包含用户 id、用户名和密码三个属性。在该类中实现空构造和带参构造，并实现每个属性的 get、set 方法，具体代码如下：

```
public class User {
        private Integer id;
        private String userName;
        private String passWord;
        public User(Integer id, String userName, String passWord) {
                super();
                this.id = id;
                this.userName = userName;
                this.passWord = passWord;
        }
        public User() {
                super();
        }
        public Integer getId() {
                return id;
        }
        public void setId(Integer id) {
                this.id = id;
        }
        public String getUserName() {
                return userName;
        }
```

```
        public void setUserName(String userName) {
                this.userName = userName;
        }
        public String getPassWord() {
                return passWord;
        }
        public void setPassWord(String passWord) {
                this.passWord = passWord;
        }
        @Override
        public String toString() {
                return "[id=" + id + ", userName=" + userName + ", passWord="
                                + passWord + "]\n";
        }
}
```

步骤四：创建 DBOpenHelper 类继承自 SQLiteOpenhelper，实现其构造、onCreate()和onUpgrade()方法，具体代码如下：

```
//定义工具类, 继承 SQLiteOpenHelper
public class DBOpenHelper extends SQLiteOpenHelper {

        // 创建对象的时候, 需要传入上下文环境
        public DBOpenHelper(Context context) {
                super(context, "user.db", null, 2);
                /*
                 * 由于父类没有无参构造函数，必须显式调用有参的构造函数
                 * 参数 1: 上下文环境，用来确定数据库文件存储的目录
                 * 参数 2: 数据库文件的名字
                 * 参数 3: 生成游标的工厂，填 null 就是使用默认的
                 * 参数 4: 数据库的版本，从 1 开始
                 */
        }

        @Override
        public void onCreate(SQLiteDatabase db) {
                System.out.println("onCreate");
                // 执行 SQL 语句, 创建表
                db.execSQL(
"CREATE TABLE user(id INTEGER PRIMARY KEY AUTOINCREMENT, username)");
        }

        @Override
        public void onUpgrade(SQLiteDatabase db, int oldVersion, int newVersion) {
                System.out.println("onUpgrade");
                // 执行 SQL 语句, 添加表
                db.execSQL("ALTER TABLE user ADD password INTEGER");
        }

}
```

步骤五：打开 MainActivity.java 文件，在该文件中实现数据库的 CRUD 操作，并通过用户交互界面完成功能的开发，具体代码如下：

```
public class MainActivity extends Activity {
    // 声明控件对象
    private EditText edtUserName, edtPassWord;
    private TextView txtShow;
    private SQLiteDatabase db;
    @Override
    protected void onCreate(Bundle savedInstanceState) {
        super.onCreate(savedInstanceState);
        setContentView(R.layout.activity_main);
        // 获取控件资源
        edtUserName = (EditText) findViewById(R.id.edtusername);
        edtPassWord = (EditText) findViewById(R.id.edtpassword);
        txtShow = (TextView) findViewById(R.id.txtshow);
        // 调用创建数据库对象
        onCreateDB();
    }

    // 实现 myOnClick 监听方法
    public void myOnClick(View view) {
        // 获取用户输入的账号和密码
        String userName = edtUserName.getText().toString();
        String passWord = edtPassWord.getText().toString();
        User user = new User();
        user.setUserName(userName);
        user.setPassWord(passWord);
        // 判断按钮
        switch (view.getId()) {
        case R.id.btninsert:
            // 调用插入方法
            insert(user);
            break;
        case R.id.btndel:
            // 调用插入方法
            delete(userName);
            break;
        case R.id.btnupdate:
            // 调用插入方法
            update(user);
            break;
        case R.id.btnquery:
            // 调用查询方法
            List<User> list = queryAll();
            txtShow.setText("");
            // 放入 TextView
            for (User user2 : list) {
```

```
                    txtShow.append(user2.toString());
                }
                break;
            }
        }

    // 创建数据库
    private void onCreateDB() {
        // 创建数据库
        SQLiteOpenHelper helper = new DBOpenHelper(this);
        // 获取数据库可写连接
        db = helper.getWritableDatabase();
    }

    // 插入方法
    public void insert(User user) {
        // 插入语句
        db.execSQL("INSERT INTO user(username, password) VALUES(?, ?)",
                new Object[] { user.getUserName(), user.getPassWord() });
        Toast.makeText(this, "插入成功！ ", Toast.LENGTH_SHORT).show();
    }

    // 删除方法
    public void delete(String userName) {
        // 删除语句
        db.execSQL("DELETE FROM user WHERE username=?",
                new Object[] { userName });
        Toast.makeText(this, "删除成功！ ", Toast.LENGTH_SHORT).show();
    }

    // 修改方法
    public void update(User user) {
        // 修改语句
        db.execSQL("UPDATE user SET password=? WHERE username=?",
                new Object[] {user.getPassWord(), user.getUserName() });
        Toast.makeText(this, "修改成功！ ", Toast.LENGTH_SHORT).show();
    }

    // 查询方法
    public List<User> queryAll() {
        // 用游标对象保存获取的值
        Cursor c = db.rawQuery("SELECT id, username, password FROM user", null);
        // 新建集合
        List<User> persons = new ArrayList<User>();
        // 循环把游标里的值放入集合
        while (c.moveToNext()) {
            User p = new User(c.getInt(0), c.getString(1), c.getString(2));
            persons.add(p);
```

```
                }
                c.close();
                Toast.makeText(this, "查询成功！ ", Toast.LENGTH_SHORT).show();
                return persons;
        }
}
```

上面的程序中 onCreateDB()方法用于创建或打开 SQLite 数据库。当用户单击程序中的“插入”按钮时，程序会调用 insert(User user)方法向底层数据表中插入一行记录，单击“删除”按钮将会删除用户输入账号对应的数据库字段，单击“修改”按钮时将会修改用户输入账号对应的数据库密码字段，单击“查询”按钮时会把所有查询的内容显示在文本框内。

5.3.4 SQLite 数据库存储位置

当程序运行后会在当前项目目录下创建“user.db”数据库，打开 File Explorer，即可看到如图 5-7 所示的界面。

Threads | Heap | Allocation T | Network Sta | File Explore | Emulator C | System Info

Name	Size	Date	Time	Permissions	Info
com.example.sqlitedemo		2015-05-28	19:53	drwxr-x--x	
cache		2015-05-28	19:42	drwxrwx--x	
databases		2015-05-28	19:44	drwxrwx--x	
user.db	20480	2015-05-28	19:55	-rw-rw----	
user.db-journal	12824	2015-05-28	19:55	-rw-------	
lib		2015-05-28	19:53	lrwxrwxrwx	-> /da

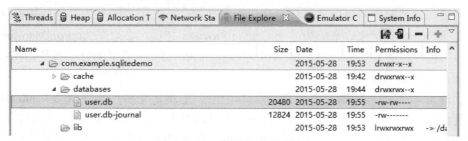

图 5-7　user.db 数据库存放位置

SQLite 数据保存在/data/data/<package name>/ databases 目录下。通过 File Explorer 面板的导出文件按钮将该 db 文件导出到桌面，使用 SQLite ExpertProfessional 工具打开该数据库文件可看到如图 5-8 所示的文件内容。

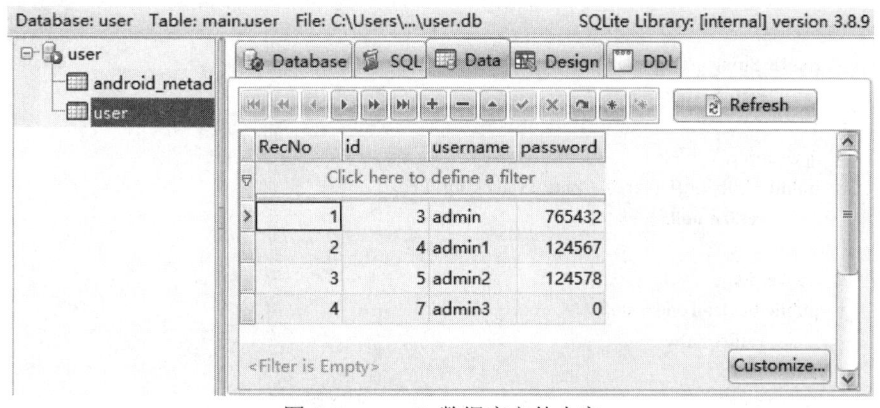

图 5-8　user.db 数据库文件内容

5.4 数据存储四：ContentProvider

Android 系统中的数据是私有的，这些私有数据包括文件数据和数据库数据以及一些其他类型的数据。Android 中的两个程序之间可以进行数据交换，此功能就是通过 ContentProvider 实现的。一个 ContentProvider 类实现了一组标准的方法接口，从而能够让其他的应用保存或读取此 ContentProvider 的各种数据类型。一个程序可以通过实现一个 ContentProvider 的抽象接口将自己的数据共享出去，其他程序看不到具体数据内容，也不用看到这个应用暴露的数据在应用当中是如何存储的，重要的是其他程序可以通过这一套标准及统一的接口和程序里的数据打交道，可以读取程序的数据，也可以删除程序的数据，在中间也会涉及一些权限的问题。

5.4.1 ContentProvider 的使用

当然要与其他的应用程序共享应用程序中的数据，首先需要通过为其他应用程序提供标准化的 ContentProvider 接口来成为一个 ContentProvider 之后，必须在 AndroidManifest.xml 文件中将应用程序注册为 ContentProvider。成为 ContentProvider 最直接的方法就是将需要共享的信息存储在 SQLite 数据库中。

（1）继承 ContentProvider。下面是一段代码框架，它提供了应用程序成为 ContentProvider 所需实现的基本接口。在下面的代码中，将会给出这五大方法的具体实现。可以使用 Eclipse 来方便地创建一个新的类继承自 ContentProvider，这时新类中会包含所需的基本覆写。代码如下：

```
public class MyProvider extends ContentProvider {
    @Override
    public int delete(Uri arg0, String arg1, String[] arg2) {
        return 0;
    }
    @Override
    public String getType(Uri arg0) {
        return null;
    }
    @Override
    public Uri insert(Uri arg0, ContentValues arg1) {
        return null;
    }
    @Override
    public boolean onCreate() {
        return false;
    }
    @Override
    public Cursor query(Uri arg0, String[] arg1, String arg2, String[] arg3,
```

```
                    String arg4) {
            return null;
        }
        @Override
        public int update(Uri arg0, ContentValues arg1, String arg2, String[] arg3)
    {
            return 0;
        }
    }
```

（2）注册 ContentProvider。打开 AndroidManifest.xml 配置文件，在<application></application>标签之间，注册 MyProvider 类，代码如下：

```
<provider
    android:name=".MyProvider"
    android:authorities=" com.example.sqlitedemo" />
```

（3）定义数据 URI。Provider 应用程序需要定义一个 URI 以供其他应用程序访问该 ContentProvider。这一 URI 的定义形式必须为 public static final Uri，名称必须为 CONTENT_URI，并且必须以 content://开头。URI 必须是唯一的，最好使用 ContentProvider 的完整有效类名称命名它。

下面的代码示例为自定义的 Provider 创建了一个 URI 名称：

```
public static final Uri CONTENT_URI
= Uri.parse("content://com.example.sqlitedemo.MyProvider/user/_id");
```

“content://”：标准前缀，用来说明一个 ContentProvider 控制这些数据，无法改变，与http://协议一样。

“com.example.sqlitedemo.MyProvider”：URI 的标识，用于唯一标识这个 ContentProvider，外部调用者可以根据这个标识来找到它。它定义了哪个 ContentProvider 提供这些数据。对于第三方应用程序，为了保证 URI 标识的唯一性，它必须是一个完整的、小写的类名。这个标识在元素的 authorities 属性中说明：一般是定义该 ContentProvider 的包、类的名称。

“user”：路径（path），通俗的讲就是要操作的数据库中表的名字，可自定义，但在使用时保持一致就可以了，例：

"content://com.xxx.provider.myprovider/tablename"

“_id”：如果 URI 中包含表示需要获取的记录的 ID，则就返回该 id 对应的数据。如果没有 ID，就表示返回全部，"content://com.xxx.provider.myprovider/tablename/#"中的#表示数据 id。

5.4.2 ContentProvider 的 CRUD 操作

【实例 5.6】SQLiteProvider 使用数据共享实现 CRUD 操作应用。

下面的程序示范了如何在Android应用中使用ContentProvider操作另外一个应用的SQLite数据库。本书提供了两个项目，分别是 SQLiteProvider 和 ContentProviderDemo。在 SQLiteProvider 项目中通过“SQLiteProvider.java”类共享 insert、delete、update 和 query 数据库操作接口，然后在 ContentProviderDemo 项目中通过“ProviderTest.java”类，调用上述提供

的接口，完成本项目对其他项目数据库的增删改查操作。

步骤一：打开 Eclipse，导入 SQLiteProvider 工程。

步骤二：找到位于“com.example.sqliteprovider.dao”包下方的“DBTest.java”类，使用 JUnit 测试执行，如图 5-9 所示，创建数据库并在数据库中插入数据，具体操作已在类中指出。

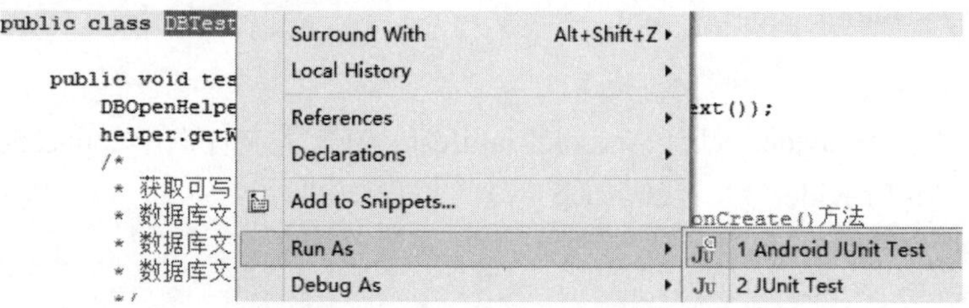

图 5-9　运行 Junit 单元测试

运行成功后，会在 JUnit Test 窗口显示绿条，如图 5-10 所示。

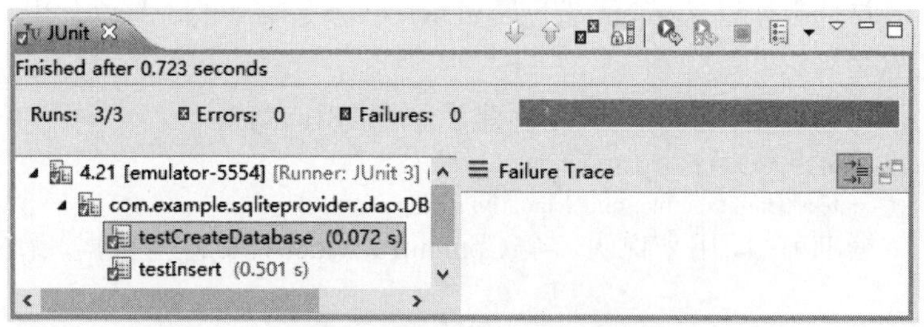

图 5-10　JUnit Test 运行成功

步骤三：把 SQLiteProvider 程序正常运行到虚拟机，主界面如图 5-11 所示。

SQLiteProvider

序号	姓名	余额
1	李四	1999
2	王五	2999
3	小强	3999
4	小明	4999
5	小龙	5999
6	张三	6999
7	小芳	7999
8	宋六	8999
9	闫七	9999
10	韩八	10000

图 5-11　主界面显示的数据库数据

步骤四：SQLiteProvider 项目共享数据库增删改查接口，具体实现见“com.example.sqliteprovider.provider”包中的“SQLiteProvider.java”，代码如下：

```
public class SQLiteProvider extends ContentProvider {
    // Uri 匹配器, 用来匹配传入的 Uri
    private UriMatcher matcher = new UriMatcher(UriMatcher.NO_MATCH);
    private DBOpenHelper helper;
    private static final int PERSON = 1;
    private static final int PERSON_ID = 2;

    // 创建内容提供者时执行
    @Override
    public boolean onCreate() {
        helper = new DBOpenHelper(getContext());
        // 设置一个 Uri, 如果匹配到 person, 返回 1
        matcher.addURI("com.example.sqliteprovider.provider", "person", PERSON);
        // 如果匹配到 student, 返回 2
        matcher.addURI("com.example.sqliteprovider.provider", "person/#", PERSON_ID);
        return true;
    }

    @Override
    public Cursor query(Uri uri, String[] projection, String selection, String[] selectionArgs, String sortOrder) {
        SQLiteDatabase db = helper.getReadableDatabase();
        // 用匹配器匹配传入的 uri
        switch (matcher.match(uri)) {
            case PERSON_ID:
                // 获取 uri 最后的 id
                long id = ContentUris.parseId(uri);
                // 构建查询条件
                selection = selection == null ? "id=" + id : selection + " AND id=" + id;
            case PERSON:
                // 执行查询
                return db.query("person", projection, selection, selectionArgs, null, null, sortOrder);
            default:
                throw new RuntimeException("Uri 不能识别: " + uri);
        }
    }

    @Override
    public Uri insert(Uri uri, ContentValues values) {
        SQLiteDatabase db = helper.getWritableDatabase();
        switch (matcher.match(uri)) {
            case PERSON:
                // 插入记录，得到 id
                long id = db.insert("person", "id", values);
                // 把 id 跟在 uri 后面返回
                return ContentUris.withAppendedId(uri, id);
```

```
                default:
                    throw new RuntimeException("Uri 不能识别: " + uri);
            }
        }

        @Override
        public int delete(Uri uri, String selection, String[] selectionArgs) {
            SQLiteDatabase db = helper.getWritableDatabase();
            // 用匹配器匹配传入的 uri
            switch (matcher.match(uri)) {
                case PERSON_ID:
                    // 获取 uri 最后的 id
                    long id = ContentUris.parseId(uri);
                    // 构建查询条件
                    selection = selection == null ? "id=" + id : selection + " AND id=" + id;
                case PERSON:
                    return db.delete("person", selection, selectionArgs);
                default:
                    throw new RuntimeException("Uri 不能识别: " + uri);
            }
        }

        @Override
        public int update(Uri uri, ContentValues values, String selection, String[] selectionArgs) {
            SQLiteDatabase db = helper.getWritableDatabase();
            // 用匹配器匹配传入的 uri
            switch (matcher.match(uri)) {
                case PERSON_ID:
                    // 获取 uri 最后的 id
                    long id = ContentUris.parseId(uri);
                    // 构建查询条件
                    selection = selection == null ? "id=" + id : selection + " AND id=" + id;
                case PERSON:
                    return db.update("person", values, selection, selectionArgs);
                default:
                    throw new RuntimeException("Uri 不能识别: " + uri);
            }
        }

        @Override
        public String getType(Uri uri) {
            return null;
        }
}
```

步骤五：打开 AndroidManifest.xml 中，注册 ContentProvider，具体代码如下：

```
<?xml version="1.0" encoding="utf-8"?>
<manifest xmlns:android="http://schemas.android.com/apk/res/android"
```

```
    package="com.example.sqliteprovider"
    android:versionCode="1"
    android:versionName="1.0" >

    <uses-sdk
        android:minSdkVersion="8"
        android:targetSdkVersion="17" />

    <instrumentation
        android:name="android.test.InstrumentationTestRunner"
        android:targetPackage="com.example.sqliteprovider" />

    <application
        android:allowBackup="true"
        android:icon="@drawable/ic_launcher"
        android:label="@string/app_name"
        android:theme="@style/AppTheme" >

         <uses-library android:name="android.test.runner" />

        <activity
            android:name=".ui.SimpleCursorAdapterActivity"
            android:label="@string/app_name" >
            <intent-filter>
                <action android:name="android.intent.action.MAIN" />

                <category android:name="android.intent.category.LAUNCHER" />
            </intent-filter>
        </activity>
          <!-- 注册 ContentProvider -->
        <provider
            android:exported="true"
            android:name=".provider.SQLiteProvider"
            android:authorities="com.example.sqliteprovider.provider" />
    </application>

</manifest>
```

到这步为止，SQLiteProvider 工程中已经共享好了相关的数据库增删改查接口，之后通过 ContentProviderDemo 工程去调用 SQLiteProvider 中共享的数据库操作接口，实现数据共享。

步骤六：打开 Eclipse 导入 ContentProviderDemo 项目。

步骤七：使用 SQLiteProvider 工程共享的 CRUD 接口，在当前 ContentProviderDemo 项目中对 SQLiteProvider 工程中的数据库进行增删改查操作。具体实现见“com.example.contentproviderdemo”包下的“ProviderTest.java”类，代码如下：

```
public class ProviderTest extends AndroidTestCase {

	public void test1() {
		// 获取解析器对象
		ContentResolver resolver = getContext().getContentResolver();

		// 访问内容提供者
		Uri uri = Uri.parse("content://com.example.sqliteprovider.provider");
		ContentValues values = new ContentValues();

		resolver.insert(uri, values);
		resolver.delete(uri, null, null);
		resolver.update(uri, values, null, null);
		resolver.query(uri, null, null, null, null);
	}

	public void testInsert() {
		ContentResolver resolver = getContext().getContentResolver();
		Uri uri = Uri.parse("content://com.example.sqliteprovider.provider/person");

		ContentValues values = new ContentValues();
		values.put("name", "provider");
		values.put("balance", 12345);
		// 插入数据，并且得到这条数据的 uri
		uri = resolver.insert(uri, values);
		System.out.println(uri);
	}

	public void testDelete() {
		ContentResolver resolver = getContext().getContentResolver();
		Uri uri = Uri.parse("content://com.example.sqliteprovider.provider/person");
		int count = resolver.delete(uri, null, null);
		System.out.println(count);
	}

	public void testUpdate() {
		ContentResolver resolver = getContext().getContentResolver();
		Uri uri = Uri.parse("content://com.example.sqliteprovider.provider/person");

		ContentValues values = new ContentValues();
		values.put("name", "update");
		values.put("balance", 54321);
		// 更新数据，得到影响记录数
		int count = resolver.update(uri, values, null, null);
		System.out.println(count);
	}

	public void testQuery() {
		ContentResolver resolver = getContext().getContentResolver();
		Uri uri = Uri.parse("content://com.example.sqliteprovider.provider/person");
```

```
                Cursor c = resolver.query(uri, new String[]{ "id", "name", "balance" },
                        "balance>?", new String[]{ "9000" }, "balance DESC");
                while (c.moveToNext()) {
                    Person p = new Person(c.getInt(0), c.getString(1), c.getInt(2));
                    System.out.println(p);
                }
        }
}
```

上述 testInsert、testDelete、testUpdate 和 testQuery 方法分别调用了 SQLiteProvider 项目中共享的 CRUD 接口，实现了当前程序调用另外一个程序的数据库。

步骤八：使用 JUnit 单元测试，运行“ProviderTest.java”类，即可实现对 SQLiteProvider 数据库的增删改查操作。

5.5　实训项目

到目前为止我们学完的 Android 中常见的数据保存方式，下面通过一个案例来巩固之前所学。

移动日记

【实例 5.7】移动日记。

本实例实现了一个基本的日记本项目，我们不但可以实现增删改查的操作，而且还能够把数据库当中的数据显示在一个 ListView 当中，通过对 ListView 的操作，实现对数据的增删改查操作。

本例提供的项目是 MobileDiary，具体实现流程如下所示。

步骤一：创建名为 MobileDiary 的项目。

步骤二：编写 res/layout/activity_main.xml 布局文件，此文件是项目执行后首先显示的界面，如图 5-12 所示。

图 5-12　移动日记主界面

具体代码实现如下：

```
<RelativeLayout xmlns:android="http://schemas.android.com/apk/res/android"
    android:layout_width="match_parent"
    android:layout_height="match_parent"
    android:background="#ffffff" >

    <TextView
        android:id="@+id/txt_title"
        android:layout_width="match_parent"
        android:layout_height="wrap_content"
        android:background="#06bc71"
        android:gravity="center"
        android:padding="4dp"
        android:shadowColor="#05814e"
        android:shadowDx="1"
        android:shadowDy="1"
        android:shadowRadius="1"
        android:text="移动日记"
        android:textColor="#ffffff"
        android:textSize="22sp" />
    <TextView
        android:id="@+id/txt_tag"
        android:layout_width="wrap_content"
        android:layout_height="wrap_content"
        android:layout_below="@+id/txt_title"
        android:layout_centerHorizontal="true"
        android:text="单击下方的按钮添加"
        android:textColor="#adaead" />

    <ListView
        android:id="@+id/android:list"
        android:layout_width="match_parent"
        android:layout_height="match_parent"
        android:background="#ffffff"
        android:layout_below="@+id/txt_tag"/>

    <Button
        android:id="@+id/btn_add"
        android:layout_width="50dp"
        android:layout_height="50dp"
        android:layout_alignParentBottom="true"
        android:layout_centerHorizontal="true"
        android:layout_marginBottom="10dp"
        android:background="@drawable/btn_edt_selecotr"
        android:onClick="myOnClick" />

</RelativeLayout>
```

步骤三：我们需要为 ListView 中每条展示的数据设置一个布局，以便更好地展示数据库内容。在 res/layout 下新建 list_item.xml 布局，具体代码如下：

```
<RelativeLayout xmlns:android="http://schemas.android.com/apk/res/android"
    android:id="@+id/row"
    android:layout_width="wrap_content"
    android:layout_height="40dp"
    android:background="#ffffff"
    android:paddingLeft="5dp"
    android:paddingRight="5dp" >

    <TextView
        android:id="@+id/text1"
        android:layout_width="wrap_content"
        android:layout_height="wrap_content"
        android:maxWidth="200dip"
        android:text="第 1 组第 1 项"
        android:textSize="20sp"
        android:layout_centerVertical="true"/>

    <TextView
        android:id="@+id/created"
        android:layout_width="wrap_content"
        android:layout_height="wrap_content"
        android:layout_alignParentRight="true"
        android:layout_marginLeft="10dip"
        android:text="2015 年 7 月 20 日"
        android:layout_centerVertical="true"
        android:gravity="center_vertical"/>
</RelativeLayout>
```

步骤四：新建 DiaryDbAdapter.java 文件，编写该文件实现数据库的基本操作。定义类 DiaryDbAdapter，此类用于封装 DatabaseHelper 和 SQLiteDatabase 类，使我们对数据库的操作更加安全和方便。定义 DatabaseHelper 类的方法非常简单，只不过在项目里边，在 onUpgrade 增加了升级的代码，具体代码如下所示：

```
public class DiaryDbAdapter {

    public static final String KEY_TITLE = "title";
    public static final String KEY_BODY = "body";
    public static final String KEY_ROWID = "_id";
    public static final String KEY_CREATED = "created";
    private static final String TAG = "DiaryDbAdapter";
    private DatabaseHelper mDbHelper;
    private SQLiteDatabase mDb;
    private static final String DATABASE_CREATE = "create table diary (_id integer primary key autoincrement, "
            + "title text not null, body text not null, created text not null);";
    private static final String DATABASE_NAME = "database";
```

```
    private static final String DATABASE_TABLE = "diary";
    private static final int DATABASE_VERSION = 1;
    private final Context mCtx;

    private static class DatabaseHelper extends SQLiteOpenHelper {

        DatabaseHelper(Context context) {
            super(context, DATABASE_NAME, null, DATABASE_VERSION);
        }
        @Override
        public void onCreate(SQLiteDatabase db) {
            db.execSQL(DATABASE_CREATE);
        }
        @Override
        public void onUpgrade(SQLiteDatabase db, int oldVersion, int newVersion)
    {
            db.execSQL("DROP TABLE IF EXISTS diary");
            onCreate(db);
        }
    }

    public DiaryDbAdapter(Context ctx) {
        this.mCtx = ctx;
    }

    public DiaryDbAdapter open() throws SQLException {
        mDbHelper = new DatabaseHelper(mCtx);
        mDb = mDbHelper.getWritableDatabase();
        return this;
    }

    public void closeclose() {
        mDbHelper.close();
    }

    public long createDiary(String title, String body) {
        ContentValues initialValues = new ContentValues();
        initialValues.put(KEY_TITLE, title);
        initialValues.put(KEY_BODY, body);
        Calendar calendar = Calendar.getInstance();
        String created = calendar.get(Calendar.YEAR) + "年"
                + calendar.get(Calendar.MONTH) + "月"
                + calendar.get(Calendar.DAY_OF_MONTH) + "日    "
                + calendar.get(Calendar.HOUR_OF_DAY) + ":"
                + calendar.get(Calendar.MINUTE);
        initialValues.put(KEY_CREATED, created);
```

```
            return mDb.insert(DATABASE_TABLE, null, initialValues);
        }

        public boolean deleteDiary(long rowId) {
            return mDb.delete(DATABASE_TABLE, KEY_ROWID + "=" + rowId, null) > 0;
        }

        public Cursor getAllNotes() {
            return mDb.query(DATABASE_TABLE, new String[] { KEY_ROWID, KEY_TITLE,
                                    KEY_BODY, KEY_CREATED }, null, null, null, null, null);
        }

        public Cursor getDiary(long rowId) throws SQLException {
            Cursor mCursor =
            mDb.query(true, DATABASE_TABLE, new String[] { KEY_ROWID, KEY_TITLE,
                        KEY_BODY, KEY_CREATED }, KEY_ROWID + "=" + rowId, null, null,
                        null, null, null);
            if (mCursor != null) {
                mCursor.moveToFirst();
            }
            return mCursor;
        }

        public boolean updateDiary(long rowId, String title, String body) {
            ContentValues args = new ContentValues();
            args.put(KEY_TITLE, title);
            args.put(KEY_BODY, body);
            Calendar calendar = Calendar.getInstance();
            String created = calendar.get(Calendar.YEAR) + "年"
                        + calendar.get(Calendar.MONTH) + "月"
                        + calendar.get(Calendar.DAY_OF_MONTH) + "日"
                        + calendar.get(Calendar.HOUR_OF_DAY) + "小时"
                        + calendar.get(Calendar.MINUTE) + "分种";
            args.put(KEY_CREATED, created);
            return mDb.update(DATABASE_TABLE, args, KEY_ROWID + "=" + rowId, null) > 0;
        }
}
```

在 DiaryDbAdapter 类中提供了如下方法：

（1）open()：调用这个方法后，如果数据库还没有建立，那么会建立数据库，如果数据库已经建立了，那么会返回可写的数据库实例。

（2）close()：调用此方法，DatabaseHelper 会关闭对数据库的访问。

（3）createDiary(String title，String body)通过一个 title 和 body 字段在数据库当中创建一条新的记录。

（4）deleteDiary (long rowId)：通过记录的 id，删除数据库中的那条记录。

（5）getAllNotes()：得到 diary 表中所有的记录，并且以一个 Cursor 的形式进行返回。

（6）getDiary (long rowId)：通过记录的主键 id，得到特定的一条记录。

（7）updateDiary (long rowId，String .title，String body)：更新主键 id 为 rowId 那条记录中的两个字段 title 和 body 字段的内容。

步骤五：打开 MainActivity.java 文件，此文件的实现流程如下所示。在代码中 myOnClick() 单击事件实现跳转到添加日记的界面，方法 onMenuItemSelected()，用于处理选择按下 Menu 时弹出的选项，具体代码如下所示：

```
public class MainActivity extends ListActivity {

        private static final int ACTIVITY_CREATE = 0;
        private static final int ACTIVITY_EDIT = 1;
        private static final int DELETE_ID = Menu.FIRST + 1;
        private DiaryDbAdapter mDbHelper;
        private Cursor mDiaryCursor;

        @Override
        public void onCreate(Bundle savedInstanceState) {
                super.onCreate(savedInstanceState);
                requestWindowFeature(Window.FEATURE_NO_TITLE);
                setContentView(R.layout.activity_main);
                mDbHelper = new DiaryDbAdapter(this);
                mDbHelper.open();
                renderListView();

        }

        private void renderListView() {
                mDiaryCursor = mDbHelper.getAllNotes();
                startManagingCursor(mDiaryCursor);
                String[] from = new String[] { DiaryDbAdapter.KEY_TITLE,
                                DiaryDbAdapter.KEY_CREATED };
                int[] to = new int[] { R.id.text1, R.id.created };
                SimpleCursorAdapter notes = new SimpleCursorAdapter(this, R.layout.list_item,
                                mDiaryCursor, from, to);
                setListAdapter(notes);
        }

        public void myOnClick(View v){
                createDiary();
        }

        @Override
        public boolean onCreateOptionsMenu(Menu menu) {
```

```
            super.onCreateOptionsMenu(menu);
            menu.add(0, DELETE_ID, 0, R.string.menu_delete);
            return true;
      }

      @Override
      public boolean onMenuItemSelected(int featureId, MenuItem item) {
            switch (item.getItemId()) {
            case DELETE_ID:
                  mDbHelper.deleteDiary(getListView().getSelectedItemId());
                  renderListView();
                  return true;
            }
            return super.onMenuItemSelected(featureId, item);
      }

      private void createDiary() {
            Intent i = new Intent(this, EditActivity.class);
            startActivityForResult(i, ACTIVITY_CREATE);
      }

      @Override
      // 需要对 position 和 id 进行一个很好的区分
      // position 指的是单击的这个 ViewItem 在当前 ListView 中的位置
      // 每一个和 ViewItem 绑定的数据，肯定都有一个 id，通过这个 id 可以找到那条数据。
      protected void onListItemClick(ListView l, View v, int position, long id) {
            super.onListItemClick(l, v, position, id);
            Cursor c = mDiaryCursor;
            c.moveToPosition(position);
            Intent i = new Intent(this, EditActivity.class);
            i.putExtra(DiaryDbAdapter.KEY_ROWID, id);
            i.putExtra(DiaryDbAdapter.KEY_TITLE,
                        c.getString(c.getColumnIndexOrThrow(DiaryDbAdapter.KEY_TITLE)));
            i.putExtra(DiaryDbAdapter.KEY_BODY,
                        c.getString(c.getColumnIndexOrThrow(DiaryDbAdapter.KEY_BODY)));
            startActivityForResult(i, ACTIVITY_EDIT);
      }

      @Override
      protected void onActivityResult(int requestCode, int resultCode,
                  Intent intent) {
            super.onActivityResult(requestCode, resultCode, intent);
            renderListView();
      }
}
```

步骤六：当单击如图 5-12 所示界面中下方的圆形编辑按钮，跳转到添加日记的界面，如图 5-13 所示。

图 5-13　添加日记界面

新建名为 EditActivity.java 的 Activity，同时会在 res/layout 目录下为该 Activity 新建名为 activity_edit.xml 布局文件，打开该布局文件，绘制如图 5-13 所示的界面，具体代码如下：

```
<LinearLayout xmlns:android="http://schemas.android.com/apk/res/android"
    xmlns:tools="http://schemas.android.com/tools"
    android:layout_width="match_parent"
    android:layout_height="match_parent"
    tools:context=".MainActivity"
    android:orientation="vertical"
    android:background="#ffffff">
    <TextView
        android:id="@+id/txt_title"
        android:layout_width="match_parent"
        android:layout_height="wrap_content"
        android:text="移动日记"
        android:textSize="22sp"
        android:textColor="#ffffff"
        android:gravity="center"
        android:padding="4dp"
        android:background="#06bc71"
        android:shadowColor="#05814e"
        android:shadowDx="1"
        android:shadowDy="1"
        android:shadowRadius="1"/>

    <LinearLayout
```

```
            android:layout_width="fill_parent"
            android:layout_height="wrap_content"
            android:orientation="vertical" >

            <TextView
                android:layout_width="wrap_content"
                android:layout_height="wrap_content"
                android:padding="2px"
                android:text="@string/title" />

            <EditText
                android:id="@+id/title"
                android:layout_width="fill_parent"
                android:layout_height="wrap_content"
                android:layout_weight="1" />
        </LinearLayout>

        <TextView
            android:layout_width="wrap_content"
            android:layout_height="wrap_content"
            android:text="@string/body" />

        <EditText
            android:id="@+id/body"
            android:layout_width="fill_parent"
            android:layout_height="fill_parent"
            android:layout_weight="1"
            android:gravity="top"
            android:scrollbars="vertical" />

        <Button
            android:id="@+id/confirm"
            android:layout_width="match_parent"
            android:layout_height="wrap_content"
            android:text="@string/confirm" />

    </LinearLayout>
```

步骤七：打开 EditActivity.java 文件，在该文件中实现往数据库添加日记的操作，具体代码如下：

```
public class EditActivity extends Activity {

    private EditText mTitleText;
    private EditText mBodyText;
    private Long mRowId;
    private DiaryDbAdapter mDbHelper;

```

```
    @Override
    protected void onCreate(Bundle savedInstanceState) {
        super.onCreate(savedInstanceState);
        mDbHelper = new DiaryDbAdapter(this);
        mDbHelper.open();
        requestWindowFeature(Window.FEATURE_NO_TITLE);
        setContentView(R.layout.activity_edit);

        mTitleText = (EditText) findViewById(R.id.title);
        mBodyText = (EditText) findViewById(R.id.body);
        Button confirmButton = (Button) findViewById(R.id.confirm);

        mRowId = null;
        // 每一个 intent 都会带一个 Bundle 型的 extras 数据。
        Bundle extras = getIntent().getExtras();
        if (extras != null) {
            String title = extras.getString(DiaryDbAdapter.KEY_TITLE);
            String body = extras.getString(DiaryDbAdapter.KEY_BODY);
            mRowId = extras.getLong(DiaryDbAdapter.KEY_ROWID);
            if (title != null) {
                mTitleText.setText(title);
            }
            if (body != null) {
                mBodyText.setText(body);
            }
        }

        confirmButton.setOnClickListener(new View.OnClickListener() {
            public void onClick(View view) {
                String title = mTitleText.getText().toString();
                String body = mBodyText.getText().toString();
                if (mRowId != null) {
                    mDbHelper.updateDiary(mRowId, title, body);
                } else
                    mDbHelper.createDiary(title, body);
                Intent mIntent = new Intent();
                setResult(RESULT_OK, mIntent);
                finish();
            }

        });
    }
}
```

步骤八：在 AVD 或手机上运行该项目，即完成了移动日记应用的开发。

5.6 本章小结

本章详细介绍了 SharedPreferences、File、SQLite 和 ContentProvider 这四种数据存储方式。当存储简单数据和配置信息时，SharedPreferences 是首选， 如果 SharedPreferences 不够用，那就需要选择文件存储或数据库存储了。在 Android 中，文件存储有两种方式，分别是存储到 ROM 和存储到 SDCard，两种文件存储的位置和实现方式略有不同。当使用 SQLite 进行数据存储，熟悉 SQLiteOpenHelper 和 SQLiteDatabase 这两个类，将会使我们事半功倍。

在 Android 中数据存储都是私有的，其他应用程序都无法访问。如果需要共享数据则需用 ContentProvider，再通过 ContentResolver 获取共享的数据。采用 SharedPreferences 共享数据，需要使用 SharedPreferences API 读写数据。采用文件方式对外共享数据，需要进行文件操作读写数据，而使用 ContentProvider 共享数据的好处是统一了数据访问方式，使数据访问变得更为快捷和规范。

在以后的开发过程中，读者可根据设计目标、性能需求、空间需求等选择合适的数据存储方式进行项目开发。

5.7 本章习题

请为移动日记项目完善删除日记的功能。

6

Android 图形图像

学习目标：

在掌握 Android 图形图像处理知识的基础上，达到能利用图片及动画增强界面美感的效果。

【知识目标】

- 掌握图片的加载及显示
- 掌握 Tween 和 Frame 动画知识
- 掌握 Android 2D 绘图知识
- 了解图形特效的制作

【技能目标】

- 能在应用程序中合理使用图片
- 能通过动画增加程序的动感
- 能绘制 2D 图形
- 能使用图形特效

6.1 图片

6.1.1 使用图片文件创建 Drawable 对象

最简单的一种方式是在工程的资源文件下面保存图片，该图片会被 Eclipse 自动在 R 类中创建引用，然后可以通过 R.drawable.imageName 使用该图片对象。下面看一个实例，实例步骤说明如下。

程序运行效果如图 6-1 所示。

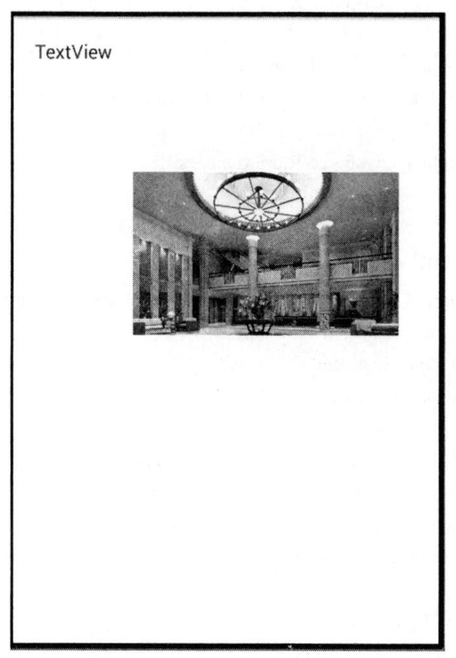

图 6-1 ImageView 的应用

【实例 6.1】图片的显示。

步骤一：将一个图片文件 myimage.jpg 放置在工程的资源文件夹 drawable 下，这里需要注意的是文件的命名规则，必须以小写字母 a～z、数字 0～9、下划线“_”和英文句号“.”符号组成，不能包含中文及其他特殊字符，否则整个工程将不能编译。

步骤二：创建布局文件 main.xml 并在其中添加一个 ImageView 组件。

步骤三：创建 Activity，并实例化 ImageView 组件对象。

步骤四：调用 ImageView 的 setImageResource()方法，引用资源 id。

Activity 具体代码如下所示：

```
package com.drawable;

import android.app.Activity;
import android.os.Bundle;
import android.widget.ImageView;

public class MainActivity extends Activity {
      //声明图片视图 ImageView
      private ImageView myImageView;
```

```

    /** Called when the activity is first created. */
    @Override
    public void onCreate(Bundle savedInstanceState) {
        super.onCreate(savedInstanceState);
        setContentView(R.layout. activity_main);
        //通过 findViewById 方法获得 ImageView
        myImageView = (ImageView) findViewById(R.id.ImageView01);
        //为 ImageView 设置图片资源
        myImageView.setImageResource(R.drawable.test);
    }
}
```

布局文件 activity_main.xml 程序代码清单如下所示。

```
<RelativeLayout xmlns:android="http://schemas.android.com/apk/res/android"
    xmlns:tools="http://schemas.android.com/tools"
    android:layout_width="match_parent"
    android:layout_height="match_parent"
    android:paddingBottom="@dimen/activity_vertical_margin"
    android:paddingLeft="@dimen/activity_horizontal_margin"
    android:paddingRight="@dimen/activity_horizontal_margin"
    android:paddingTop="@dimen/activity_vertical_margin"
    tools:context="com.example.helloandroid.MainActivity" >
    <ImageView
        android:id="@+id/ImageView01"
        android:layout_width="wrap_content"
        android:layout_height="wrap_content"
        android:layout_alignParentLeft="true"
        android:layout_alignParentTop="true"
        android:layout_marginLeft="74dp"
        android:layout_marginTop="98dp"
        android:src="@drawable/ic_launcher" />
    <TextView
        android:id="@+id/textView1"
        android:layout_width="wrap_content"
        android:layout_height="wrap_content"
        android:layout_alignParentTop="true"
        android:layout_alignRight="@+id/imageView1"
        android:text="TextView" />
</RelativeLayout>
```


6.1.2 使用 XML 文件定义 Drawable 属性

在布局文件中通过定义组件的 Drawable 属性，为组件添加背景或显示图片，从而使界面美观。例如，可以在布局文件中定义图片按钮的图片及应用程序的图标等。

【实例 6.2】通过 Drawable 属性显示图片。

所下的代码演示了如何在 AndroidManifest.xml 配置文件中引用资源图标。

AndroidManifest.xml 详细代码如下所示：

```
<?xml version="1.0" encoding="utf-8"?>
<manifest xmlns:android="http://schemas.android.com/apk/res/android"
        package="cn.com.farsight.drawable"
        android:versionCode="1"
        android:versionName="1.0">

<application android:icon="@drawable/icon"
                        android:label="@string/app_name">
    <activity android:name="com.MainActivity"
                    android:label="@string/app_name">
        <intent-filter>
            <action android:name="android.intent.action.MAIN" />
            <category android:name="android.intent.category.LAUNCHER" />
        </intent-filter>
    </activity>
</application>
<uses-sdk android:minSdkVersion="8"/>
</manifest>
```

在上一个例子中也可以通过这种方式来配置 ImageView 的图片资源。代码如下所示：

```
<ImageView
        android:id="@+id/ImageView01"
        android:layout_width="wrap_content"
        android:layout_height="wrap_content"
        android:src="@drawable/test"      >
</ImageView>
```

6.1.3 Bitmap 和 BitmapFactory

除了以上提到的两种图片保存以外，还可以将图片文件保存在 SDCard 中，那么此时如何访问保存在 SDCard 中的图片文件呢？在这种情况下，可以通过 Bitmap 和 BitmapFactory 两个类来读取文件。下面的代码演示了如何从 SDCard 或 drawable 文件夹中读取图片文件并将其设置为壁纸。效果如图 6-2 所示。

【实例 6.3】修改手机桌面壁纸。

程序步骤说明如下：

步骤一：在 SDCard 中添加一个名称为 wallpaper.jgp 的图片文件。

步骤二：创建 Activity。

步骤三：在 AndroidManifest.xml 添加用户权限“SET_WALLPAPER”，从而使应用程序可以设置桌面图片。

图 6-2 修改手机壁纸

在 AndroidManifest.xml 添加用户权限“READ_EXTERNAL_STORAGE”，从而使应用程序可以读取 SD 上的文件。

步骤四：在 Activity 的 onCreate()方法中通过 BitmapFactory 的 decodeFile()方法传递文件路径，获取 Bitmap 对象。

步骤五：用 setWallpaper()方法设置桌面。

MainActivity.java 程序代码清单如下所示。

```
import java.io.IOException;
import android.app.Activity;
import android.graphics.Bitmap;
import android.graphics.BitmapFactory;
import android.os.Bundle;
public class MainActivity extends Activity {
@Override
public void onCreate(Bundle savedInstanceState) {
super.onCreate(savedInstanceState);
setContentView(R.layout. activity_main);
        //图片路径，实际运行中注意查看图片的完整路径，可能不是以/sdcard 开头
String path="/sdcard/wallpaper.jpg";
//通过 BitmapFactory 获取 Bitmap 实例
Bitmap bm = BitmapFactory.decodeFile(path);
//可以利用 drawable 下的已有图片
// Bitmap bm = BitmapFactory.decodeResource(this.getResources(),R.drawable.test);
        try{
```

```
//设置桌面
setWallpaper(bm);
}catch(IOException e){
e.printStackTrace();
  }
}
  }
获取 SD 卡路径的常用方法如下：
//判断 sd 卡是否存在
boolean sdCardExist = Environment.getExternalStorageState()
                                         .equals(Android.os.Environment.MEDIA_MOUNTED);
if (sdCardExist)
{
      sdDir = Environment.getExternalStorageDirectory();//获取 SD 卡的根目录
}
```

6.2 动画

Android 平台提供了两类动画。一类是 Tween 动画，就是对场景里的对象不断地进行图像变化来产生动画效果（旋转、平移、放缩和渐变）；另一类是 Frame 动画，即顺序地播放事先做好的图像，与 gif 图片原理类似。

6.2.1 Tween 动画

Tween 动画又称“补间动画”“中间动画”。Tween 动画在 Android 中分为四类，它们分别是：

（1）AlphaAnimation（透明度动画）

（2）TranslateAnimation（平移动画）

（3）ScaleAnimation（缩放动画）

（4）RotateAnimation（旋转动画）

这四个类都继承自 android.view.Animation 类，用来表示从状态 A 向状态 B 变化的一个过程，所以英文名字叫 Tween 动画，中文名叫“补间动画”“中间动画”。Tween 动画有两种实现方式：

（1）通过 java 源代码实现。

（2）通过 xml 配置文件实现。

AlphaAnimation（透明度动画），有两个构造函数，分别是：

AlphaAnimation(Context context, AttributeSet attrs)：第二个参数是一个属性集，之后会详细对 AttributeSet 进行讲解；

AlphaAnimation(float fromAlpha, float toAlpha)：第一个参数是初始透明度，第二个参数是

终止透明度。

TranslateAnimation（平移动画）有三个构造函数，分别是：

- TranslateAnimation(Context context, AttributeSet attrs)；
- TranslateAnimation(float fromXDelta, float toXDelta, float fromYDelta, float toYDelta)，分别对应 x 轴的起始、终点坐标，以及 y 轴的起始、终点坐标；
- TranslateAnimation(int fromXType, float fromXValue, int toXType, float toXValue, int fromYType, float fromYValue, int toYType, float toYValue)，第一个参数是 x 轴方向的值的参照（Animation. ABSOLUTE、Animation.RELATIVE_TO_SELF 或 Animation.RELATIVE_TO_PARENT）；第二个参数是第一个参数类型的起始值；第三个参数与第四个参数是 x 轴方向的终点参照与对应值。

如果全部选择 Animation.ABSOLUTE，其实就是第二个构造函数。

以 x 轴为例介绍参照与对应值的关系，如果选择参照为 Animation.ABSOLUTE，那么对应的值应该是具体的坐标值，比如 100～300，指绝对的屏幕像素单位。

如果选择参照为 Animation.RELATIVE_TO_SELF 或者 Animation.RELATIVE_TO_PARENT 指的是相对于自身或父控件，对应值应该理解为相对于自身或者父控件的几倍或百分之多少。

ScaleAnimation（缩放动画）有四个构造函数，分别是：

- ScaleAnimation(Context context, AttributeSet attrs)；
- ScaleAnimation(float fromX, float toX, float fromY, float toY)，同 TranslateAnimation (float fromXDelta, float toXDelta, float fromYDelta, float toYDelta)；
- ScaleAnimation(float fromX, float toX, float fromY, float toY, float pivotX, float pivotY)，这里解释后面两个参数，pivot 英文意思为“枢轴”，也就是支点。通过这两个参数可以控制缩放动画的方向，这个点不会随对象大小变化而变化；
- ScaleAnimation(float fromX, float toX, float fromY, float toY, int pivotXType, float pivotXValue, int pivotYType, float pivotYValue)，如果理解了前面所讲的，这个就不做过多说明；如果不清楚，请回头多用代码试试。

RotateAnimation（旋转动画）同样有四个构造函数，分别是：

- RotateAnimation(Context context, AttributeSet attrs)；
- RotateAnimation(float fromDegrees, float toDegrees)，第一个参数是 X 轴顺时针转动到 fromDegrees 为旋转的起始点，第二个参数是 X 轴顺时针转动到 toDegrees 为旋转的起始点。
- RotateAnimation(float fromDegrees, float toDegrees, float pivotX, float pivotY)，第三个参数 pivotX 为距离左侧的偏移量，第四个参数 pivotY 为距离顶部的偏移量。即为相对于 View 左上角(0,0)的坐标点。
- RotateAnimation(float fromDegrees, float toDegrees, int pivotXType, float pivotXValue, int pivotYType, float pivotYValue)。

下面的示例演示了各种动画的运行效果，如图 6-3 所示。

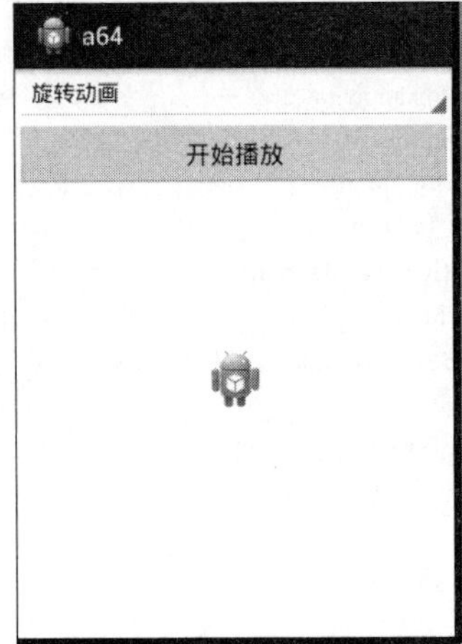

图 6-3　动画效果演示

【实例 6.4】各种 Tween 动画演示。

详细程序代码如下：

```
package com.example.a64;
import android.app.Activity;
import android.os.Bundle;
import android.view.View;
import android.view.View.OnClickListener;
import android.view.animation.AlphaAnimation;
import android.view.animation.Animation;
import android.view.animation.RotateAnimation;
import android.view.animation.ScaleAnimation;
import android.view.animation.TranslateAnimation;
import android.widget.ArrayAdapter;
import android.widget.Button;
import android.widget.ImageView;
import android.widget.Spinner;

public class TestTweenAnimation extends Activity {
//定义开始按钮
private Button start = null;
//定义动画类型下拉列表
private Spinner select = null;
//这张图片是动画执行者
private ImageView img = null;
//定义动画
private Animation tAnimation = null;
//定义一个 String 数组用于构造下拉列表的适配器
private String str[] = {
```

```
"平移动画", "透明度动画", "旋转动画", "缩放动画"
};

@Override
public void onCreate(Bundle savedInstanceState) {
      super.onCreate(savedInstanceState);
      setContentView(R.layout.activity_main);
      //分别从 xml 文件中得到每个控件
      start = (Button) findViewById(R.id.startButton);
      select = (Spinner) findViewById(R.id.select);
      img = (ImageView) findViewById(R.id.img);
      //实例化适配器
      ArrayAdapter<String> adapter = new ArrayAdapter<String>(this, android.R.layout.simple_spinner_item, str);
       select.setAdapter(adapter);
       //为开始按钮设置监听
start.setOnClickListener(new OnClickListener() {

@Override
public void onClick(View v) {
InitialAnimation();
img.startAnimation(tAnimation);
}
});
}


//初始化动画
public void InitialAnimation(){
      switch(select.getSelectedItemPosition()){
      case 0:
      tAnimation = new TranslateAnimation(0, 300, 50, 50);
//tAnimation = new TranslateAnimation(Animation.RELATIVE_TO_SELF, 0.0f,
Animation.RELATIVE_TO_SELF, 1.0f, Animation.RELATIVE_TO_PARENT, -0.5f,
Animation.RELATIVE_TO_PARENT, -0.5f);
      break;
      case 1:tAnimation = new AlphaAnimation(0.1f, 1.0f);
      break;
      case 2:tAnimation = new RotateAnimation(0.0f, +360.0f);
      break;
      case 3:
//tAnimation = new ScaleAnimation(0.0f, 1.0f, 0.0f, 1.0f);
      tAnimation = new ScaleAnimation(0.0f, 1.0f, 0.0f, 1.0f, 200.0f, 0.0f);
      break;
      }

      //为动画设置完成所需时间
      tAnimation.setDuration(2000);
      }
```

```
}
这里是 activity_main.xml
<?xml version="1.0" encoding="utf-8"?>
<RelativeLayout xmlns:android="http://schemas.android.com/apk/res/android"
    android:layout_width="fill_parent"
    android:layout_height="fill_parent"
    >
    <Spinner
    android:id="@+id/select"
    android:layout_width="fill_parent"
    android:layout_height="wrap_content"
    />
    <Button
    android:id="@+id/startButton"
    android:layout_width="fill_parent"
    android:layout_height="wrap_content"
    android:layout_below="@id/select"
    android:text="开始播放"
    />
    <ImageView
    android:id="@+id/img"
    android:layout_width="wrap_content"
    android:layout_height="wrap_content"
    android:layout_alignParentBottom="true"
    android:src="@drawable/sun"
    />
</RelativeLayout>
```

6.2.2 Frame 动画

Frame 动画是一系列图片按照一定的顺序展示的过程，和放电影的机制很相似，我们称为逐帧动画。Frame 动画可被定义在 XML 文件中，也可以完全编码实现。

如果定义在 XML 文件中，则可以放置在/res 下的 anim 或 drawable 目录中（/res/[anim | drawable]/filename.xml），文件名可以作为资源 ID 在代码中引用；如果完全由编码实现，我们需要使用到 AnimationDrawable 对象。

如果将动画定义在 XML 文件中，语法如下：

```
<?xml version="1.0" encoding="utf-8"?>
<animation-list xmlns:android="http://schemas.android.com/apk/res/android"
    android:oneshot=["true" | "false"] >
    <item          android:drawable="@[package:]drawable/drawable_resource_name"
        android:duration="integer" />
</animation-list>
```

需要注意的是：

<animation-list>元素是必须的，并且必须要作为根元素，可以包含一个或多个<item>元素；

android:onshot 如果定义为 true 的话，此动画只会执行一次，如果为 false 则一直循环。

<item>元素代表一帧动画，android:drawable 指定此帧动画所对应的图片资源，android:druation 代表此帧持续的时间，为一个整数，单位为毫秒。

新建一个名为 a65 的工程，将四张连续的图片分别命名为 f1.png、f2.png、f3.png、f4.png，放于 drawable 目录，然后新建一个 frame.xml 文件：

```
<?xml version="1.0" encoding="utf-8"?>
<animation-list xmlns:android="http://schemas.android.com/apk/res/android"
      android:oneshot="false">
    <item android:drawable="@drawable/f1" android:duration="300" />
    <item android:drawable="@drawable/f2" android:duration="300" />
    <item android:drawable="@drawable/f3" android:duration="300" />
    <item android:drawable="@drawable/f4" android:duration="300" />
</animation-list>
```

将 frame.xml 文件放置于 drawable 或 anim 目录，官方文档上是放到了 drawable 中了，大家可以根据喜好来放置，放在这两个目录都是可以运行的。

然后介绍一下布局文件 res/layout/activity_main.xml：

```
<?xml version="1.0" encoding="utf-8"?>
<LinearLayout
   xmlns:android="http://schemas.android.com/apk/res/android"
   android:orientation="vertical"
   android:layout_width="fill_parent"
   android:layout_height="fill_parent">
   <ImageView
      android:id="@+id/frame_image"
      android:layout_width="fill_parent"
      android:layout_height="fill_parent"
      android:layout_weight="1"/>
   <Button
      android:layout_width="fill_parent"
      android:layout_height="wrap_content"
      android:text="stopFrame"
      android:onClick="stopFrame"/>
   <Button
      android:layout_width="fill_parent"
      android:layout_height="wrap_content"
      android:text="runFrame"
      android:onClick="runFrame"/>
</LinearLayout>
```

定义一个 ImageView 作为动画的载体，然后定义了两个按钮，分别是停止和启动动画。

接下来介绍一下如何通过加载动画定义文件来实现动画的效果。我们首先这样写：

```
import android.app.Activity;
import android.graphics.drawable.AnimationDrawable;
import android.graphics.drawable.Drawable;
import android.os.Bundle;
```

```
import android.view.View;
import android.widget.ImageView;

public class FrameActivity extends Activity {

    private ImageView image;

    @Override
    protected void onCreate(Bundle savedInstanceState) {
        super.onCreate(savedInstanceState);
        setContentView(R.layout.frame);
        image = (ImageView) findViewById(R.id.frame_image);

        image.setBackgroundResource(R.anim.frame);
        AnimationDrawable anim = (AnimationDrawable) image.getBackground();
        anim.start();
    }
}
```

看似十分完美，跟官方文档上写的一样，然而当我们运行这个程序时会发现，它只停留在第一帧，并没有出现我们期望的动画，之后把相应的代码放在一个按钮的单击事件中，动画就顺利执行了，再移回到 onCreate 中，还是没效果，什么原因呢？如何解决呢？

出现这种现象是因为当我们在 onCreate 中调用 AnimationDrawable 的 start 方法时，窗口 Window 对象还没有完全初始化，AnimationDrawable 不能完全追加到窗口 Window 对象中，那么该怎么办呢？我们需要把这段代码放在 onWindowFocusChanged 方法中，当 Activity 展示给用户时，onWindowFocusChanged 方法就会被调用，我们正是在这个时候实现动画效果。当然，onWindowFocusChanged 是在 onCreate 之后被调用的，如图 6-4 所示。

图 6-4　Frame 动画效果演示

【实例 6.5】Frame 动画演示。

然后我们需要重写一下代码：

```
import android.app.Activity;
import android.graphics.drawable.AnimationDrawable;
import android.graphics.drawable.Drawable;
import android.os.Bundle;
import android.view.View;
import android.widget.ImageView;

public class FrameActivity extends Activity {

    private ImageView image;

    @Override
    protected void onCreate(Bundle savedInstanceState) {
        super.onCreate(savedInstanceState);
        setContentView(R.layout.frame);
        image = (ImageView) findViewById(R.id.frame_image);
    }

    @Override
    public void onWindowFocusChanged(boolean hasFocus) {
        super.onWindowFocusChanged(hasFocus);
        image.setBackgroundResource(R.anim.frame);
        AnimationDrawable anim = (AnimationDrawable) image.getBackground();
        anim.start();
    }
}
```

运行一下，动画就可以正常显示了。

如果在有些场合，我们需要用纯代码方式实现一个动画，我们可以这样写：

```
AnimationDrawable anim = new AnimationDrawable();
for (int i = 1; i <= 4; i++) {
    int id = getResources().getIdentifier("f" + i, "drawable", getPackageName());
    Drawable drawable = getResources().getDrawable(id);
    anim.addFrame(drawable, 300);
}
anim.setOneShot(false);
image.setBackgroundDrawable(anim);
anim.start();
完整的 FrameActivity.java 代码如下所示：

import android.app.Activity;
import android.graphics.drawable.AnimationDrawable;
```

```
import android.graphics.drawable.Drawable;
import android.os.Bundle;
import android.view.View;
import android.widget.ImageView;

public class FrameActivity extends Activity {

    private ImageView image;

    @Override
    protected void onCreate(Bundle savedInstanceState) {
        super.onCreate(savedInstanceState);
        setContentView(R.layout.frame);
        image = (ImageView) findViewById(R.id.frame_image);
    }

    @Override
    public void onWindowFocusChanged(boolean hasFocus) {
        super.onWindowFocusChanged(hasFocus);
        //将动画资源文件设置为 ImageView 的背景
        image.setBackgroundResource(R.anim.frame);
        //获取 ImageView 背景，此时已被编译成 AnimationDrawable
        AnimationDrawable anim = (AnimationDrawable) image.getBackground();
        //开始动画
        anim.start();
    }

    public void stopFrame(View view) {
        AnimationDrawable anim = (AnimationDrawable) image.getBackground();
        if (anim.isRunning()) { //如果正在运行,就停止
            anim.stop();
        }
    }

    public void runFrame(View view) {
        //完全编码实现的动画效果
        AnimationDrawable anim = new AnimationDrawable();
        for (int i = 1; i <= 4; i++) {
            //根据资源名称和目录获取 R.java 中对应的资源 ID
            int id = getResources().getIdentifier("f" + i, "drawable", getPackageName());
            //根据资源 ID 获取到 Drawable 对象
            Drawable drawable = getResources().getDrawable(id);
            //将此帧添加到 AnimationDrawable 中
            anim.addFrame(drawable, 300);
        }
```

```
            anim.setOneShot(false); //设置为 loop
            image.setBackgroundDrawable(anim);   //将动画设置为 ImageView 背景
            anim.start();    //开始动画
        }
    }
```

6.3 动态图形绘制

6.3.1 动态图形绘制类简介

如何动态绘制图形呢？首先来看在 Android 中如何绘制图形。其实，在 Android 中涉及的这些工具类都很形象。想象一下真正地画一张画需要哪些东西呢？首先需要一张画布，这里就是 Android 中的 Canvas；其次还需要画笔，这里就是 Android 中的 Paint；再次需要不同的颜色，这里就是 Android 中的 Color。接下来如果要画线还需要连接路径，这里就是 Android 中的 Path。还可以借助工具直接画出各种图形，如圆、椭圆、矩形等，这里就是 Android 中的 ShapeDrawable 类，当然它还有很多子类，例如 OvalShape（椭圆）、RectShape（矩形）等。

1. Canvas

Canvas 就是我们所说的画布，位于 android.graphics 包中，提供了一些画各种图形的方法，例如矩形、圆、椭圆等。该类的详细方法见表 6-1。

表 6-1 Canvas 常用方法

方法名称	方法描述
drawText(String text,float x,float y,Paint paint)	以(x,y)为起始坐标，使用 paint 绘制文本
drawPoint(float x,float y,Paint paint)	在坐标(x,y)上使用 paint 画点
drawLine(float startX,float startY,float stopX,float stopY,Paint paint)	以(startX, startY)为起始坐标点，(stopX, stopY)为终止坐标点，使用 paint 画线
drawCircle(float cx,float cy,float radius,Paint paint)	以(cx, cy)为原点，radius 为半径，使用 paint 画圆
drawOval(RectF oval,Paint paint)	使用 paint 画矩形 oval 的内切椭圆
DrawRect(RectF rect,Paint paint)	使用 paint 画矩形 rect
drawRoundRect(RectF rect,float rx,float ry,Paint paint)	画圆角矩形
clipRect(float left,float top,float right,float botton)	剪辑矩形
clipRegion(Region region)	剪辑区域

2. Paint

Paint 就是涂料的意思，用来描述图形的颜色和风格，如线宽、颜色、字体等信息。Paint 位于 android.graphics 包中，该类的详细方法见表 6-2。

表 6-2 Paint 常用方法

方法名称	方法描述
Paint()	构造方法，使用默认设置
setColor(int color)	设置颜色
setStrokeWidth(float width)	设置线宽
setTextAlign(Paint.Align align)	设置文字对齐
setTextSize(float textSize)	设置文字尺寸
setShader(Shader shader)	设置渐变
setAlpha(int a)	设置 Alpha 值
reset()	复位 Paint 默认设置

3. Color

Color 类定义了一些颜色变量和一些创建颜色的方法。颜色的定义一般使用 RGB 三原色定义。Color 位于 android.graphics 包中，其常用属性和方法见表 6-3。

表 6-3 Color 常用属性和方法

属性名称	方法描述
BLACK	黑色
BLUE	蓝色
CYAN	青色
DKGRAY	深灰色
GRAY	灰色
GREEN	绿色
LTGRAY	浅灰色
MAGENTA	品红色
RED	红色
TRANSPARENT	透明
WHITE	白色
YELLOW	黄色

4. Path

当我们想要画一个圆的时候，我们只需要指定圆心（点）和半径就可以了。那么，如果要画一个梯形呢？这里我们需要有点和连线。Path 一般用来从一点移动到另一个点的连线。Path 位于 android.graphics 包中。详细方法见表 6-4。

表 6-4 Path 常用方法

方法名称	方法描述
lineTo(float x,float y)	从最后点到指定点划线
moveTo(float x,float y)	移动到指定点
reset()	复位

6.3.2 动态图形绘制的基本思路

动态图形绘制的基本思路是，创建一个类继承 View 类（或者继承 SurfaceView 类）。覆盖 onCreate()方法，使用 Canvas 对象在界面上绘制不同的图形，使用 invalidate()方法刷新界面。下面通过一个弹球实例来讲述动态图形绘制的基本思路。该实例是在界面上动态绘制一个小球，当小球触顶或者触底时自动改变方向继续运行。运行效果如图 6-5 所示。

图 6-5 运动的小球

【实例 6.6】运动的小球。

实例步骤说明如下：

步骤一：创建一个 Android 工程，入口 Activity 的名称为 MainActivity。MainActivity 代码如下所示。

```
package com.example.a66;

import android.app.Activity;
import android.content.Context;
```

```
import android.graphics.Canvas;
import android.graphics.Color;
import android.graphics.Paint;
import android.os.Bundle;
import android.os.Handler;
import android.os.Message;
import android.util.AttributeSet;
import android.view.View;

public class MainActivity extends Activity {
/** Called when the activity is first created. */
@Override
public void onCreate(Bundle savedInstanceState) {
    super.onCreate(savedInstanceState);
}
}
```

步骤二：在 MainActivity 类中创建一个 MyView 内部类，该类实现 Runnable 接口支持多线程。

在 onDraw()方法中，定义 Paint 画笔并设置画笔颜色，使用 Canvas 的 drawCircle()方法画圆。

定义一个 update()方法，用于更新 Y 坐标。

定义一个消息处理器类 RefreshHandler，它继承 Handler 并覆盖 handleMessage()方法，在该方法中处理消息。

在线程的 run()方法中设置并发送消息。

在构造方法中启动线程。

内部类 MyView 代码如下所示：

```
class MyView extends View implements Runnable{
        //图形当前坐标
        private int x=20,y=20;
        //构造方法
        public MyView(Context context, AttributeSet attrs) {
            super(context, attrs);
            // TODO Auto-generated constructor stub
            //获得焦点
            setFocusable(true);
            //启动线程
            new Thread(this).start();
        }

        @Override
        public void run() {
            // TODO Auto-generated method stub
        while(!Thread.currentThread().isInterrupted()){
                //通过发送消息更新界面
```

```
                    Message m = new Message();
                    m.what = 0x101;
                    mRedrawHandler.sendMessage(m);
                    try {
                        Thread.sleep(100);
                    } catch (InterruptedException e) {
                        // TODO Auto-generated catch block
                        e.printStackTrace();
                    }
                }
            }

            @Override
            protected void onDraw(Canvas canvas) {
                // TODO Auto-generated method stub
                super.onDraw(canvas);
                //实例化画笔
                Paint p = new Paint();
                //设置画笔颜色
                p.setColor(Color.GREEN);
                //画图
                canvas.drawCircle(x, y, 10, p);
            }

            //更新坐标
            private void update(){
                int h=getHeight();
                y+=5;
                if(y>=h){
                    y=20;
                }
            }
    }
```

步骤三：在 MainActivity 类的 onCreate()方法中实例化 MyView 类，并将其设置为 Activity 的内容视图。代码清单对应的 onCreate()方法如下。

```
//更新界面处理器
    class RefreshHandler extends Handler{
            @Override
            public void handleMessage(Message msg) {
                // TODO Auto-generated method stub
                if(msg.what==0x101){
                    v.update();
                    v.invalidate();
                }
                super.handleMessage(msg);
            }
        }
```

```
        RefreshHandler mRedrawHandler = new RefreshHandler();
        MyView v ;
         @Override
         protected void onCreate(Bundle savedInstanceState) {
                super.onCreate(savedInstanceState);
                v = new MyView(this,null);
            setContentView(v);
         }
```

6.3.3　绘制几何图形

以前的文章里面画的都是一些矩形，今天就看看怎么在 Android 手机屏幕上绘制一些几何图形，如三角形、多边形、椭圆、圆形、正方形等。并且设置空心、实心。下面我们先来看看在 Android 中可以绘制出哪些几何图形，具体绘制图形的方法详见表 6-5。

表 6-5　绘制图形的方法

方法	说明
drawRect	绘制矩形
drawCircle	绘制圆形
drawOval	绘制椭圆
drawPath	绘制任意多边形
drawLine	绘制直线
drawPoin	绘制点

我们先看看运行效果，如图 6-6 所示。

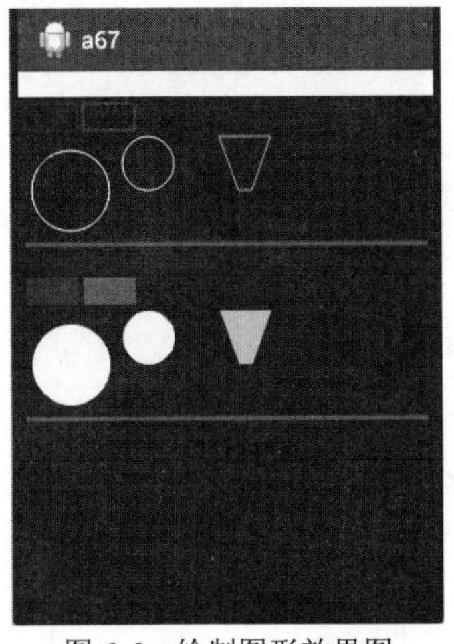

图 6-6　绘制图形效果图

【实例 6.7】绘制各种几何图形。

下面我们看例子代码。

布局文件：

```
<?xml version="1.0" encoding="utf-8"?>
<LinearLayout xmlns:android="http://schemas.android.com/apk/res/android"
      android:orientation="vertical"
      android:layout_width="fill_parent"
      android:layout_height="fill_parent"
      >
<TextView
      android:layout_width="fill_parent"
      android:layout_height="wrap_content"
      android:text="@string/hello"
      android:textColor="#00FF00"
      />
< com.example.test.GameView
android:layout_width="wrap_content"
android:layout_height="wrap_content"
/>
</LinearLayout>
```

我们自己实现的 GameView 类继承了 View，并把它作为布局文件的一部分加载了进来。

Activity01.java 代码比较简单，具体如下：

```
package com.example.test;

import android.app.Activity;
import android.os.Bundle;

public class Activity01 extends Activity {
      @Override
      protected void onCreate(Bundle savedInstanceState) {
            super.onCreate(savedInstanceState);
            setContentView(R.layout.activity_main);
      }
}
```

GameView.java 代码具体如下：

```
package com.example.a67;

import android.content.Context;
import android.graphics.Canvas;
import android.graphics.Color;
import android.graphics.Paint;
import android.graphics.Path;
import android.graphics.Rect;
```

```
import android.graphics.RectF;
import android.util.AttributeSet;
import android.view.View;

public class GameView extends View {
    // 声明 Paint 对象
    private Paint mPaint = null;

    public GameView(Context context, AttributeSet attr){
        super(context,attr);
            // 构建画笔对象
        mPaint = new Paint();
     }

    @Override
    protected void onDraw(Canvas canvas) {
        super.onDraw(canvas);

        // 设置画布为黑色背景
        canvas.drawColor(Color.BLACK);

        // 取消锯齿
        mPaint.setAntiAlias(true);

        // 设置画笔风格为空心
        mPaint.setStyle(Paint.Style.STROKE);

        {
            // 定义矩形对象
            Rect rect1 = new Rect();
            // 设置矩形大小
            rect1.left = 5;
            rect1.top = 5;
            rect1.bottom = 25;
            rect1.right = 45;

            mPaint.setColor(Color.BLUE);
            // 绘制矩形
            canvas.drawRect(rect1, mPaint);

            mPaint.setColor(Color.RED);
            // 绘制矩形
            canvas.drawRect(50, 5, 90, 25, mPaint);

            mPaint.setColor(Color.YELLOW);
```

```
        // 绘制圆形
        // 40、70 分别是圆心的 X 和 Y 坐标，30 为半径，mPaint 为画笔对象
        canvas.drawCircle(40, 70, 30, mPaint);

        // 定义椭圆
        RectF rectf1 = new RectF();
        rectf1.left = 80;
        rectf1.top = 30;
        rectf1.right = 120;
        rectf1.bottom = 70;

        mPaint.setColor(Color.LTGRAY);
        // 绘制椭圆
        canvas.drawOval(rectf1, mPaint);

        // 绘制多边形
        Path path1 = new Path();

        /**
         * 这个多变形是从起点找点，一个点、一个点的连线
         */
        path1.moveTo(150 + 5, 80 - 50); // 此点为多边形的起点
        path1.lineTo(150 + 45, 80 - 50);
        path1.lineTo(150 + 30, 120 - 50);
        path1.lineTo(150 + 20, 120 - 50);
        // 使这些点构成封闭的多边形
        path1.close();

        mPaint.setColor(Color.GRAY);
        // 绘制这个多边形
        canvas.drawPath(path1, mPaint);

        mPaint.setColor(Color.RED);
        // 设置画笔空心边框的宽度
        mPaint.setStrokeWidth(3);
        // 绘制直线
        // 这个绘制直线的方法 前 2 个参数是起点坐标，后 2 个参数是终点坐标
          我们可看出两个点的 Y 坐标都一样的
        canvas.drawLine(5, 110, 315, 110, mPaint);
    }

    // 绘制实心几何体
    // 将画笔设置为实心
    mPaint.setStyle(Paint.Style.FILL);
```

```
        {
            // 定义矩形
            Rect rect1 = new Rect();
            rect1.left = 5;
            rect1.top = 130 + 5;
            rect1.bottom = 130 + 25;
            rect1.right = 45;

            mPaint.setColor(Color.BLUE);
            // 绘制矩形
            canvas.drawRect(rect1, mPaint);

            mPaint.setColor(Color.RED);
            // 绘制矩形
            canvas.drawRect(50, 130 + 5, 90, 130 + 25, mPaint);

            mPaint.setColor(Color.YELLOW);
            // 绘制圆形，注意参数的含义，参考前面的定义
            canvas.drawCircle(40, 130 + 70, 30, mPaint);

            // 定义椭圆对象
            RectF rectf1 = new RectF();
            // 设置椭圆大小
            rectf1.left = 80;
            rectf1.top = 130+30;
            rectf1.right = 120;
            rectf1.bottom = 130 + 70;

            mPaint.setColor(Color.LTGRAY);
            // 绘制椭圆
            canvas.drawOval(rectf1, mPaint);

            // 绘制多边形
            Path path1 = new Path();

            // 设置多边形的点
            path1.moveTo(150+5, 130+80-50);
            path1.lineTo(150+45, 130+80-50);
            path1.lineTo(150+30, 130+120-50);
            path1.lineTo(150+20, 130+120-50);
            // 使这些点构成封闭的多边形
            path1.close();
            mPaint.setColor(Color.GRAY);
            // 绘制这个多边形
```

```
                canvas.drawPath(path1, mPaint);

                mPaint.setColor(Color.RED);
                mPaint.setStrokeWidth(3);
                // 绘制直线
                canvas.drawLine(5, 130 + 110, 315, 130 + 110, mPaint);
            }
        }
    }
```

6.4 图形特效

6.4.1 使用 Matrix 实现旋转、缩放和平移

在 Android 图形 API 中提供了一个 Matrix 矩形类，该类具有一个 3×3 的矩阵坐标。通过该类可以实现图形的旋转、平移和缩放。该类的详细方法见表 6-6。

表 6-6 Matrix 常用方法

方法名称	方法描述
void reset()	重置一个 matrix 对象
void set(Matrix src)	复制一个源矩阵，与构造方法 Matrix(Matrix src)一样
boolean isIdentity()	返回这个矩阵是否定义
void setRotate(float degrees)	指定一个角度以(0,0)为坐标进行旋转
void setRotate(float degrees,float px, float py)	指定一个角度以(px,py)为坐标进行旋转
void serScale(float sx, float sy)	缩放处理
void serScale(float sx, float sy,float px, float py)	以坐标(px,py)进行缩放
void setTranslate(float dx, float dy)	平移
void setSkew(float kx, float ky)	倾斜处理
void setSkew(float kx, float ky, float px, float py)	以坐标(px,py)进行倾斜

下面通过一个实例来演示 Matrix 的具体应用，在本实例中我们自定义一个 View 类，在该类中拥有一个 Bitmap 和 Matrix 实例，Bitmap 实例从系统资源加载一张图片，覆盖 View 类的 onDraw()方法，在该方法中通过 reset()方法初始化 Matrix，并设置其旋转或缩放属性，使用 Canvas 的 drawBitmap()方法将 Bitmap 重新绘制在视图中。通过键盘事件 onKeyDown()实现旋转属性和缩放属性的改变，调用 postInvalidate()方法重新绘制 Bitmap。

【实例 6.8】随手指拖动的图片。

实例步骤说明如下：

（1）新建一个项目 DragAndZoom，并准备一张照片放在 res/drawable-hdpi 目录下，如图 6-7 所示：

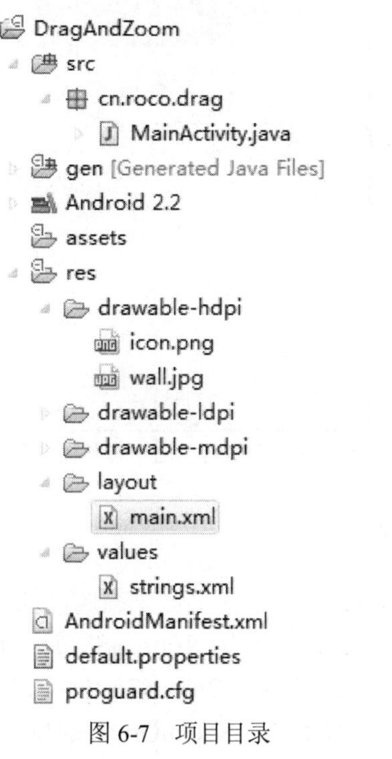

图 6-7　项目目录

（2）设置应用的 UI 界面，在 main.xml 中设置，代码如下所示：

```
<?xml version="1.0" encoding="utf-8"?>
<LinearLayout xmlns:android="http://schemas.android.com/apk/res/android"
    android:orientation="vertical"
    android:layout_width="fill_parent"
    android:layout_height="fill_parent"
    >
<ImageView
    android:layout_width="fill_parent"
    android:layout_height="wrap_content"
    android:src="@drawable/wall"
    android:id="@+id/imageView"
    android:scaleType="matrix"
    />  <!-- 指定为 matrix 类型 -->
</LinearLayout>
```

（3）具体的效果如图 6-8 所示。

图 6-8 实现可以随手指移动的图

6.4.2 使用 Shader 类渲染图形

在 Android 中，提供了 Shader 类专门用来渲染图像以及一些几何图形。

Shader 类包括了 5 个直接子类，分别为：BitmapShader、ComposeShader、LinearGradient、RadialGradient 以及 SweepGradient。其中，BitmapShader 用于图像渲染；ComposeShader 用于混合渲染；LinearGradient 用于线性渲染；RadialGradient 用于环形渲染；而 SweepGradient 则用于梯度渲染。

使用 Shader 类进行图像渲染时，首先需要构建 Shader 对象，然后通过 Paint 的 setShader() 方法来设置渲染对象，最后将这个 Paint 对象绘制到屏幕上即可。

有一点需要注意，使用不同的方式渲染图像时需要构建不同的对象。

1. BitmapShader（图像渲染）

BitmapShader 的作用是使用一张位图作为纹理来对某一区域进行填充。可以想象成在一块区域内铺瓷砖，只是这里的瓷砖是一张张位图而已。

BitmapShader 函数原型为：

public BitmapShader (Bitmap bitmap, Shader.TileMode tileX, Shader.TileMode tileY);

其中，参数 bitmap 表示用来作为纹理填充的位图；参数 tileX 表示在位图 X 方向上位图衔接形式；参数 tileY 表示在位图 Y 方向上位图衔接形式。

Shader.TileMode 有 3 种参数可供选择，分别为 CLAMP、REPEAT 和 MIRROR。

CLAMP 的作用是如果渲染器超出原始边界范围，则会复制边缘颜色对超出范围的区域进行着色。REPEAT 的作用是在横向和纵向上以平铺的形式重复渲染位图。MIRROR 的作用是在横向和纵向上以镜像的方式重复渲染位图。

2. LinearGradient（线性渲染）

LinearGradient 的作用是实现某一区域内颜色的线性渐变效果。

LinearGradient 的函数原型为：

```
public LinearGradient (float x0, float y0, float x1, float y1, int[] colors, float[] positions, Shader.TileMode tile);
```

其中，参数 x0 表示渐变的起始点 x 坐标；参数 y0 表示渐变的起始点 y 坐标；参数 x1 表示渐变的终点 x 坐标；参数 y1 表示渐变的终点 y 坐标；参数 colors 表示渐变的颜色数组；参数 positions 用来指定颜色数组的相对位置；参数 tile 表示平铺方式。

通常，参数 positions 设为 null，表示颜色数组以斜坡线的形式均匀分布。

3. ComposeShader（混合渲染）

ComposeShader 的作用是实现渲染效果的叠加，如 BitmapShader 与 LinearGradient 的混合渲染效果等。

ComposeShader 的函数原型为：

```
public ComposeShader (Shader shaderA, Shader shaderB, PorterDuff.Mode mode);
```

其中，参数 shaderA 表示某一种渲染效果；参数 shaderB 也表示某一种渲染效果；参数 mode 表示两种渲染效果的叠加模式。

PorterDuff.Mode 有 16 种参数可供选择，分别为：CLEAR、SRC、DST、SRC_OVER、DST_OVER、SRC_IN、DST_IN、SRC_OUT、DST_OUT、SRC_ATOP、DST_ATOP、XOR、DARKEN、LIGHTEN、MULTIPLY、SCREEN。

4. RadialGradient（环形渲染）

RadialGradient 的作用是在某一区域内实现环形的渐变效果。

RadialGradient 的函数原型为：

```
public RadialGradient (float x, float y, float radius, int[] colors, float[] positions, Shader.TileMode tile);
```

其中，参数 x 表示环形的圆心 x 坐标；参数 y 表示环形的圆心 y 坐标；参数 radius 表示环形的半径；参数 colors 表示环形渐变的颜色数组；参数 positions 用来指定颜色数组的相对位置；参数 tile 表示平铺的方式。

5. SweepGradient（梯度渲染）

SweepGradient 也称为扫描渲染，是指在某一中心以 x 轴正方向逆时针旋转一周而形成的扫描效果的渲染形式。

SweepGradient 的函数原型为：

```
public SweepGradient (float cx, float cy, int[] colors, float[] positions);
```

其中，参数 cx 表示扫描中心 x 的坐标；参数 cy 表示扫描中心 y 的坐标；参数 colors 表示梯度渐变的颜色数组；参数 positions 用来指定颜色数组的相对位置。

实例

在本实例中，分别实现了上述的 5 种渲染效果，如图 6-9 所示。其中，最上面的是 BitmapShader 效果图；第二排的左边是 LinearGradient 的效果图；第二排的右边是 RadialGradient

的效果图；第三排的左边是 ComposeShader 的效果图（LinearGradient 与 RadialGradient 的混合效果）；第三排的右边是 SweepGradient 的效果图。

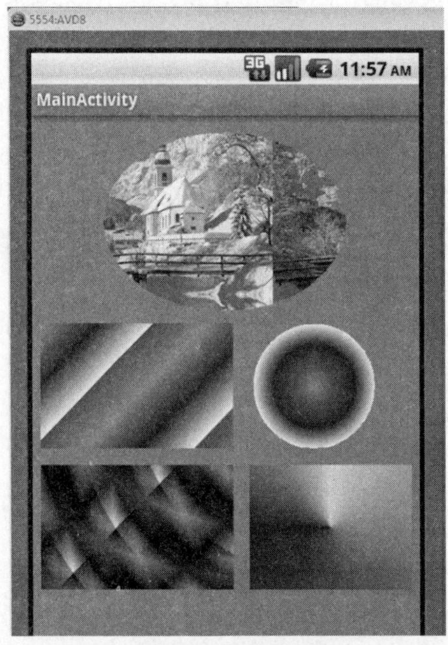

图 6-9　渲染效果

【实例 6.9】渲染效果。

定义了 MyView 类，用来绘制各种渲染效果，MyView.java 源码如下：

```
package com.example.a69;
import android.annotation.SuppressLint;
import android.content.Context;
import android.graphics.Bitmap;
import android.graphics.BitmapShader;
import android.graphics.Canvas;
import android.graphics.Color;
import android.graphics.ComposeShader;
import android.graphics.LinearGradient;
import android.graphics.Paint;
import android.graphics.PorterDuff;
import android.graphics.RadialGradient;
import android.graphics.RectF;
import android.graphics.Shader;
import android.graphics.SweepGradient;
import android.graphics.drawable.BitmapDrawable;
import android.view.View;

@SuppressLint({ "DrawAllocation", "DrawAllocation", "DrawAllocation" })
```

```
public class MyView extends View    {

    Bitmap mBitmap = null;                          //Bitmap 对象
    Shader mBitmapShader = null;                    //Bitmap 渲染对象
    Shader mLinearGradient = null;                  //线性渐变渲染对象
    Shader mComposeShader = null;                   //混合渲染对象
    Shader mRadialGradient = null;                  //环形渲染对象
    Shader mSweepGradient = null;                   //梯度渲染对象

    public MyView(Context context) {
        super(context);

        //加载图像资源
        mBitmap =BitmapFactory.decodeResource(this.getResources(),R.drawable.ic_launcher);

        //创建 Bitmap 渲染对象
        mBitmapShader = new BitmapShader(mBitmap, Shader.TileMode.REPEAT, Shader.TileMode.MIRROR);

        //创建线性渲染对象
        int mColorLinear[] = {Color.RED, Color.GREEN, Color.BLUE, Color.WHITE};
        mLinearGradient = new LinearGradient(0, 0, 100, 100, mColorLinear, null,
                Shader.TileMode.REPEAT);

        //创建环形渲染对象
        int mColorRadial[] = {Color.GREEN, Color.RED, Color.BLUE, Color.WHITE};
        mRadialGradient = new RadialGradient(350, 325, 75, mColorRadial, null, Shader.TileMode.REPEAT);

        //创建混合渲染对象
        mComposeShader = new ComposeShader(mLinearGradient, mRadialGradient,
                PorterDuff.Mode.DARKEN);

        //创建梯形渲染对象
        int mColorSweep[] = {Color.GREEN, Color.RED, Color.BLUE, Color.YELLOW, Color.GREEN};
        mSweepGradient = new SweepGradient(370, 495, mColorSweep, null);
    }

    public void onDraw(Canvas canvas) {
        super.onDraw(canvas);

        Paint mPaint = new Paint();
        canvas.drawColor(Color.GRAY);          //背景置为灰色

        //绘制 Bitmap 渲染的椭圆
        mPaint.setShader(mBitmapShader);
        canvas.drawOval(new RectF(90, 20, 90+mBitmap.getWidth(),20+mBitmap.getHeight()), mPaint);

        //绘制线性渐变的矩形
        mPaint.setShader(mLinearGradient);
```

```
            canvas.drawRect(10, 250, 250, 400, mPaint);

            //绘制环形渐变的圆
            mPaint.setShader(mRadialGradient);
            canvas.drawCircle(350, 325, 75, mPaint);

            //绘制混合渐变(线性与环形混合)的矩形
            mPaint.setShader(mComposeShader);
            canvas.drawRect(10, 420, 250, 570, mPaint);

            //绘制梯形渐变的矩形
            mPaint.setShader(mSweepGradient);
            canvas.drawRect(270, 420, 470, 570, mPaint);
        }
    }
```

在 Activity 中实例化 MyView，并通过 setContentView 加载，可以看到上述效果。

6.5 实训项目

飞针气球

“飞针气球”项目的内容源自大家在日常生活中经常参与的飞针扎气球游戏：在公园广场等公共场所，经常会有利用飞针扎气球来让小朋友参与的游戏，根据气球的数量，给予一定的奖励。同时，随着现代生活工作节奏的加快，上班一族也需要一种放松心情的小游戏，而通过针扎气球游戏，看到一个个破裂的气球及爆炸的声响，使身心得到一定的放松，愉悦心情。本款游戏以娱乐形式来放松自我，这成为了我们决定设计这款 3G 智能手机游戏软件的主要原因。最终效果如图 6-10、图 6-11 所示。

图 6-10　主界面

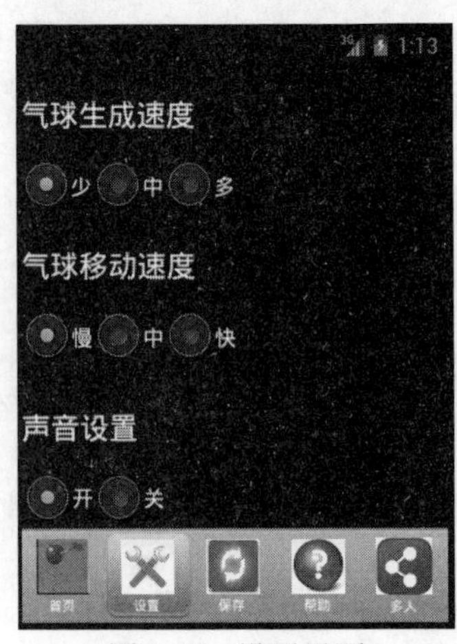

图 6-11　设置界面

6.6 本章小结

本章主要讲述了 Android 2D 图形图像的知识，包括加载和显示图片的方法，动画效果的实现，动态图形的绘制和特效等，并通过丰富的案例展示了各种效果。

6.7 本章习题

1．简述 Android 动画实现的两种方式，并说明其原理。
2．Android 图形 API 中提供了一个 Matrix 矩形类的工作机制及原理。
3．在屏幕上绘制 sin 曲线。
4．实现在屏幕上自动弹跳的小球效果。

7 综合案例开发——简易通讯录

学习目标：

本章主要巩固之前课程所学的基础知识点，从使用控件、布局进行界面设计，到使用 SQLite、Intent、Adapter 等对象完成后台开发。我们将通过本章学习掌握项目开发的整个流程。

【知识目标】

- 巩固之前课程所学的基础知识点

【技能目标】

- 灵活运用相关控件及对象完成项目开发

在每个人的手机中必不可少的软件之一就是通讯录。通讯录可以和很多功能整合到一起，如短信、电话等，形成一个集中化管理的软件。下面要介绍的通讯录是最简单的但却是通讯录功能的核心。通过之前所学到的内容，完成简易通讯录的开发。

7.1 界面设计

本章要实现的通讯录功能主要有：联系人的添加、删除、编辑、查看，向选中的联系人打电话、发信息等。程序首先呈现在我们眼前的是主界面，在主界面中以一个 ListView 显示当前添加的所有通讯录，如图 7-1 所示。

添加联系人和编辑联系人的界面，如图 7-2 所示。

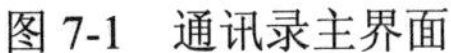

图 7-1　通讯录主界面

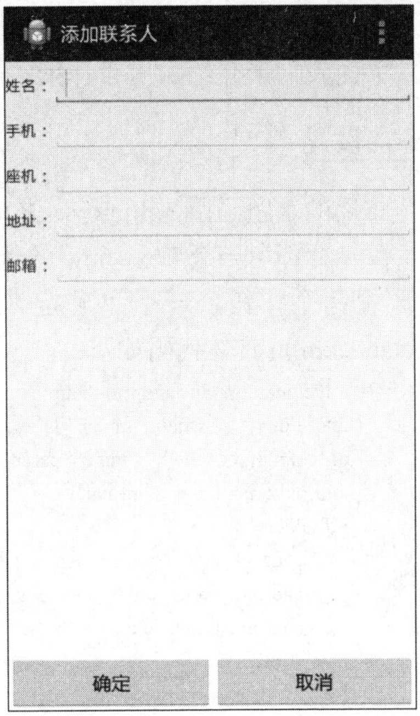

图 7-2　添加联系人界面

单击主页面中任意一个名字，即可查看对应联系人的详细信息，如图 7-3 所示。

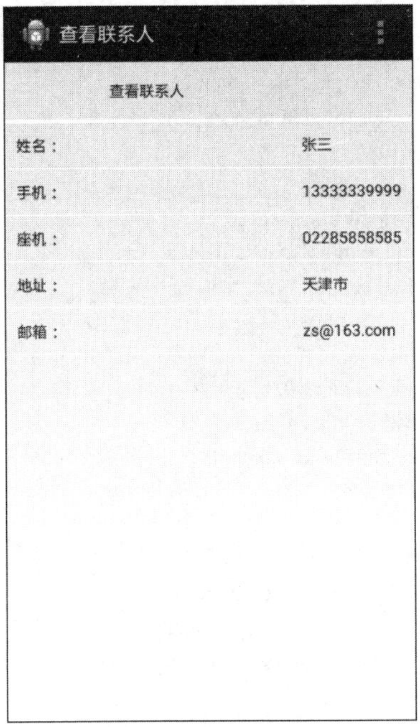

图 7-3　查看联系人界面

以上就是本程序需要呈现的三个不同的界面，第一个界面不需要布局文件，直接在程序中实现。下面我们来介绍如何设计第二个界面，也就是添加联系人的布局文件。

7.1.1 布局设置

整个布局的主题采用 TableRow 来呈现，采用 TableRow 的好处是对齐很方便，适合于表格布局。每一行都用一个 TextView 的标签加一个 EditText 用于提供用户填写信息，最后在布局底部放置两个按钮，一个“确定”按钮，一个“取消”按钮。文件位置为 res/layout/editorcontacts.xml，具体代码如下：

```
<RelativeLayout xmlns:android="http://schemas.android.com/apk/res/android"
    android:layout_width="match_parent"
    android:layout_height="match_parent"
    android:orientation="vertical" >
    <TableLayout
        android:layout_width="match_parent"
        android:layout_height="wrap_content"
        android:stretchColumns="1" >
        <TableRow
            android:id="@+id/TableRow01"
            android:layout_width="match_parent"
            android:layout_height="wrap_content" >
            <!-- 姓名标签 -->
            <TextView
                android:id="@+id/TextView01"
                android:layout_width="wrap_content"
                android:layout_height="wrap_content"
                android:text="@string/name" />
            <!-- 姓名 -->
            <EditText
                android:id="@+id/EditText01"
                android:layout_width="match_parent"
                android:layout_height="wrap_content" />
        </TableRow>
        <TableRow
            android:id="@+id/TableRow02"
            android:layout_width="match_parent"
            android:layout_height="wrap_content" >
            <!-- 手机标签 -->
            <TextView
                android:id="@+id/TextView02"
                android:layout_width="wrap_content"
                android:layout_height="wrap_content"
                android:text="@string/mobile" />
            <!-- 手机 -->
            <EditText
                android:id="@+id/EditText02"
```

```
                android:layout_width="match_parent"
                android:layout_height="wrap_content"
                android:inputType="phone" />
        </TableRow>
        <TableRow
            android:id="@+id/TableRow03"
            android:layout_width="match_parent"
            android:layout_height="wrap_content" >
            <!-- 座机标签 -->
            <TextView
                android:id="@+id/TextView03"
                android:layout_width="wrap_content"
                android:layout_height="wrap_content"
                android:text="@string/home" />
            <!-- 座机电话 -->
            <EditText
                android:id="@+id/EditText03"
                android:layout_width="match_parent"
                android:layout_height="wrap_content"
                android:inputType="phone" />
        </TableRow>
        <TableRow
            android:id="@+id/TableRow04"
            android:layout_width="match_parent"
            android:layout_height="wrap_content" >
            <!-- 住址标签 -->
            <TextView
                android:id="@+id/TextView04"
                android:layout_width="wrap_content"
                android:layout_height="wrap_content"
                android:text="@string/address" />
            <!-- 住址 -->
            <EditText
                android:id="@+id/EditText04"
                android:layout_width="match_parent"
                android:layout_height="wrap_content" />
        </TableRow>
        <TableRow
            android:id="@+id/TableRow05"
            android:layout_width="match_parent"
            android:layout_height="wrap_content" >
            <!-- 电子邮箱标签 -->
            <TextView
                android:id="@+id/TextView05"
                android:layout_width="wrap_content"
                android:layout_height="wrap_content"
                android:text="@string/email" />
            <!-- 电子邮箱 -->
```

```
                <EditText
                    android:id="@+id/EditText05"
                    android:layout_width="match_parent"
                    android:layout_height="wrap_content"
                    android:inputType="textEmailAddress" />
            </TableRow>
        </TableLayout>
        <LinearLayout
            android:layout_width="match_parent"
            android:layout_height="wrap_content"
            android:layout_alignParentBottom="true"
            android:orientation="horizontal" >
            <!-- 保存按钮 -->
            <Button
                android:id="@+id/Button01"
                android:layout_width="wrap_content"
                android:layout_height="wrap_content"
                android:layout_weight="1"
                android:text="@string/ok" />
            <!-- 取消按钮 -->
            <Button
                android:id="@+id/Button02"
                android:layout_width="wrap_content"
                android:layout_height="wrap_content"
                android:layout_weight="1"
                android:text="@string/cancel" />
        </LinearLayout>
</RelativeLayout>
```

7.1.2 添加“查看联系人”页面

查看联系人的页面与添加页面很类似，基本就是把显示的 EditView 控件换成了 TextView 控件。文件位置为 res/layout/viewuser.xml，具体代码如下：

```
<TableLayout xmlns:android="http://schemas.android.com/apk/res/android"
    android:layout_width="fill_parent"
    android:layout_height="fill_parent"
    android:stretchColumns="1" >
    <TableRow>
        <TextView
            android:layout_column="1"
            android:gravity="center_horizontal"
            android:padding="15dip"
            android:text="查看联系人" />
    </TableRow>
    <!-- 间隔线 -->
    <View
```

```
        android:layout_height="3dip"
        android:background="#ffffffff" />
    <TableRow>
        <!-- 姓名标签 -->
        <TextView
            android:layout_column="1"
            android:padding="10dip"
            android:text="姓名：" />
        <!-- 姓名 -->
        <TextView
            android:id="@+id/TextView_Name"
            android:padding="10dip" />
    </TableRow>
    <View
        android:layout_height="1dip"
        android:background="#ffffffff" />
    <TableRow>
        <!-- 手机标签 -->
        <TextView
            android:layout_column="1"
            android:padding="10dip"
            android:text="手机：" />
        <!-- 手机 -->
        <TextView
            android:id="@+id/TextView_Mobile"
            android:padding="10dip" />
    </TableRow>
    <View
        android:layout_height="1dip"
        android:background="#ffffffff" />
    <TableRow>
        <!-- 座机标签 -->
        <TextView
            android:layout_column="1"
            android:padding="10dip"
            android:text="座机：" />
        <!-- 座机 -->
        <TextView
            android:id="@+id/TextView_Home"
            android:padding="10dip" />
    </TableRow>
    <View
        android:layout_height="1dip"
        android:background="#ffffffff" />
    <TableRow>
        <!-- 住址标签 -->
        <TextView
            android:layout_column="1"
```

```
                        android:padding="10dip"
                        android:text="地址：" />
                    <!-- 住址 -->
                    <TextView
                        android:id="@+id/TextView_Address"
                        android:padding="10dip" />
                </TableRow>
                <View
                    android:layout_height="1dip"
                    android:background="#ffffffff" />
                <TableRow>
                    <!-- 电子邮箱标签 -->
                    <TextView
                        android:layout_column="1"
                        android:padding="10dip"
                        android:text="邮箱：" />
                    <!-- 电子邮箱 -->
                    <TextView
                        android:id="@+id/TextView_Email"
                        android:padding="10dip" />
                </TableRow>
                <View
                    android:layout_height="1dip"
                    android:background="#ffffffff" />
            </TableLayout>
```

7.2 功能实现

界面设计完后，下面开始实现通讯录的具体功能。

7.2.1 创建数据库

因为通讯录的信息需要保存在数据库中，所以首先要创建一个数据库，新建 DBHelper.java，具体代码如下所示。创建一个数据库，数据库文件名称为 mycontacts.db，数据表名称为 contacts，数据表包含的字段有“姓名”“手机”“座机”“地址”和“电子邮箱”。新建 DBHelper 继承自 SQLiteOpenHelper，具体代码如下：

```
public class DBHelper extends SQLiteOpenHelper {
        // 数据库名
        public static final String DATABASE_NAME = "mycontacts.db";
        // 版本
        public static final int DATABASE_VERSION = 2;
        // 表名
        public static final String CONTACTS_TABLE = "contacts";
        // 创建表
        private static final String DATABASE_CREATE = "CREATE TABLE "
```

```
                    + CONTACTS_TABLE + " (" + ContactColumn._ID
                    + " integer primary key autoincrement," + ContactColumn.NAME
                    + " text," + ContactColumn.MOBILENUM + " text,"
                    + ContactColumn.HOMENUM + " text," + ContactColumn.ADDRESS
                    + " text," + ContactColumn.EMAIL + " text )";

        public DBHelper(Context context) {
            super(context, DATABASE_NAME, null, DATABASE_VERSION);
        }

        // 创建数据库
        public void onCreate(SQLiteDatabase db) {
            db.execSQL(DATABASE_CREATE);
        }

        // 升级
        public void onUpgrade(SQLiteDatabase db, int oldVersion, int newVersion)
        {
            db.execSQL("DROP TABLE IF EXISTS " + CONTACTS_TABLE);
            onCreate(db);
        }
}
```

7.2.2 创建 ContactColumn 类

新建一个类 ContactColumn，专门用于存放数据表的列名和索引值，具体代码如下所示：

```
//定义数据
public class ContactColumn implements BaseColumns {
        public ContactColumn() {
        }

        // 列名
        public static final String NAME = "name"; // 姓名
        public static final String MOBILENUM = "mobileNumber";// 移动电话
        public static final String HOMENUM = "homeNumber"; // 座机
        public static final String ADDRESS = "address"; // 地址
        public static final String EMAIL = "email"; // 邮箱
        // 列索引值
        public static final int _ID_COLUMN = 0;
        public static final int NAME_COLUMN = 1;
        public static final int MOBILENUM_COLUMN = 2;
        public static final int HOMENUM_COLUMN = 3;
        public static final int ADDRESS_COLUMN = 4;
        public static final int EMAIL_COLUMN = 5;

        // 查询结果
        public static final String[] PROJECTION = { _ID, NAME, MOBILENUM, HOMENUM,ADDRESS, EMAIL };
}
```

7.2.3 为数据库提供操作类

创建了数据库后，下一步需要为数据库提供操作类。新建 ContactsProvider.java 类继承自 ContentProvider。代码 8～12 行设置了 ContentProvider 用于操作的 URI 地址，这些 URI 地址可以直接定位到指定的数据。代码 13～25 行定义了 URI 的匹配类型，有 CONTACTS 和 CONTACT_ID 两种，其中“#”号表示匹配任意字符，因此 CONTACTS 表示匹配到所有联系人信息，而 CONTACTS_ID 则表示对指定 id 的联系人进行操作。

在 onCreate 中初始化一些变量，接下来主要实现 insert()、update()、delete()、query()这几个函数。在 delete()函数中，首先通过 uirMatcher.match(uri)对传递过来的 URI 地址进行匹配，分情况进行处理，最后要通知更改并返回删除的个数。Insert 函数也类似，先判断 URI 地址是否合法，这时候 URI 地址必须匹配 CONTACTS，接着实例化一个 ContentValues，并将参数的值传递进去，若没有相应的值，则该值设置为“”（空）。如果成功插入数据，则返回该数据的 URI 地址。在 query 函数中新建一个 SQLiteQueryBuilder 变量用于执行查询，若传递的是 CONTACTS_ID，则将相应的判断条件加入 where 语句，如代码 145～146 行所示，最后通过 qb.query 执行操作查询，返回游标。Update 操作与 delete()函数类似，只不过操作的函数从 delete 变成了 update。

最后代码 65～79 行实现了 getType()函数，主要用来根据 uri 地址返回相应的 MIME 类型，以调用对应的 Activity。具体代码如下所示。

```
1   public class ContactsProvider extends ContentProvider {
2       //标签
3       private static final String TAG= "ContactsProvider";
4       //数据库帮助类
5       private DBHelper dbHelper;
6       //数据库
7       private SQLiteDatabase contactsDB;
8       //数据库操作 uri 地址
9       public static final String AUTHORITY = com.example.provider.ContactsProvider";
10      public static final String CONTACTS_TABLE = "contacts";
11      public static final Uri CONTENT_URI = Uri.parse("content://" + AUTHORITY + "/"+CONTACTS_TABLE);
12
13      //下面是自定义的类型
14      public static final int CONTACTS = 1;
15      public static final int CONTACT_ID = 2;
16      private static final UriMatcher uriMatcher;
17      static
18      {
19          //没有匹配的信息
20          uriMatcher = new UriMatcher(UriMatcher.NO_MATCH);
21          //全部联系人信息
22          uriMatcher.addURI(AUTHORITY,"contacts",CONTACTS);
23          //单独一个联系人信息
```

```
		uriMatcher.addURI(AUTHORITY,"contacts/#",CONTACT_ID);
	}
	//取得数据库
	@Override
	public boolean onCreate()
	{
		dbHelper = new DBHelper(getContext());
		//执行创建数据库
		contactsDB = dbHelper.getWritableDatabase();
		return (contactsDB == null) ? false : true;
	}

	// 删除指定数据列
	@Override
	public int delete(Uri uri, String where, String[] selectionArgs)
	{
		int count;
		switch (uriMatcher.match(uri))
		{
			//删除满足条件 where 的行
			case CONTACTS:
				count = contactsDB.delete(CONTACTS_TABLE, where, selectionArgs);
				break;
			case CONTACT_ID:
				//取得联系人的 id 信息
				String contactID = uri.getPathSegments().get(1);
				//删除满足 where 条件，并且 id 值为 contactID 的记录
				count = contactsDB.delete(CONTACTS_TABLE,
								ContactColumn._ID
								+ "=" + contactID
								+ (!TextUtils.isEmpty(where) ? " AND ("
								+ where + ")" : ""),selectionArgs);
				break;
			default:
				throw new IllegalArgumentException("Unsupported URI: " + uri);
		}
		getContext().getContentResolver().notifyChange(uri, null);
		return count;
	}

	// URI 类型转换
	public String getType(Uri uri)
	{
		switch (uriMatcher.match(uri))
		{
			//所有联系人
			case CONTACTS:
				return "vnd.android.cursor.dir/vnd.guo.android.mycontacts";
```

```
                //指定联系人
                case CONTACT_ID:
                    return "vnd.android.cursor.item/vnd.guo.android.mycontacts";
                default:
                    throw new IllegalArgumentException("Unsupported URI: " + uri);
            }
    }

    // 插入数据
    public Uri insert(Uri uri, ContentValues initialValues)
    {
        //判断 URI 地址是否合法
        if (uriMatcher.match(uri) != CONTACTS)
        {
            throw new IllegalArgumentException("Unknown URI " + uri);
        }
        ContentValues values;
        if (initialValues != null)
        {
            values = new ContentValues(initialValues);
            Log.e(TAG + "insert", "initialValues is not null");
        }
        else
        {
            values = new ContentValues();
        }
        // 如果对应的名称没有值，则设置默认值为""
        if (values.containsKey(ContactColumn.NAME) == false)
        {
            values.put(ContactColumn.NAME, "");
        }
        if (values.containsKey(ContactColumn.MOBILENUM) == false)
        {
            values.put(ContactColumn.MOBILENUM, "");
        }
        if (values.containsKey(ContactColumn.HOMENUM) == false)
        {
            values.put(ContactColumn.HOMENUM, "");
        }
        if (values.containsKey(ContactColumn.ADDRESS) == false)
        {
            values.put(ContactColumn.ADDRESS, "");
        }
        if (values.containsKey(ContactColumn.EMAIL) == false)
        {
            values.put(ContactColumn.EMAIL, "");
        }
        Log.e(TAG + "insert", values.toString());
```

```
            //插入数据
            long rowId = contactsDB.insert(CONTACTS_TABLE, null, values);
            if (rowId > 0)
            {
                //将 id 值加入 URI 地址中
                Uri noteUri = ContentUris.withAppendedId(CONTENT_URI, rowId);
                //通知改变
                getContext().getContentResolver().notifyChange(noteUri, null);
                Log.e(TAG + "insert", noteUri.toString());
                return noteUri;
            }
            throw new SQLException("Failed to insert row into " + uri);
        }

        // 查询数据
        public Cursor query(Uri uri, String[] projection, String selection, String[] selectionArgs, String sortOrder){
            Log.e(TAG + ":query", " in Query");
            SQLiteQueryBuilder qb = new SQLiteQueryBuilder();
            //设置要查询的数据表
            qb.setTables(CONTACTS_TABLE);

            switch (uriMatcher.match(uri))
            {
                //构建 where 语句，定位到指定 id 值的列
                case CONTACT_ID:
                    qb.appendWhere(ContactColumn._ID + "=" + uri.getPathSegments().get(1));
                    break;
                default:
                    break;
            }
            //查询
            Cursor c = qb.query(contactsDB, projection, selection, selectionArgs, null, null, sortOrder);
            //设置通知改变的 URI
            c.setNotificationUri(getContext().getContentResolver(), uri);
            return c;
        }

        // 更新数据库
        public int update(Uri uri, ContentValues values, String where, String[] selectionArgs) {
            int count;
            Log.e(TAG + "update", values.toString());
            Log.e(TAG + "update", uri.toString());
            Log.e(TAG + "update :match", "" + uriMatcher.match(uri));
            switch (uriMatcher.match(uri))
            {
                //根据 where 条件批量进行更新
                case CONTACTS:
                    Log.e(TAG + "update", CONTACTS + "");
```

```
                    count = contactsDB.update(CONTACTS_TABLE, values, where, selectionArgs);
                    break;
                //更新指定行
                case CONTACT_ID:
                    String contactID = uri.getPathSegments().get(1);
                    Log.e(TAG + "update", contactID + "");
                    count = contactsDB.update(CONTACTS_TABLE, values, ContactColumn._ID + "="
                    + contactID + (!TextUtils.isEmpty(where) ? " AND (" + where + ")" : ""), selectionArgs);
                    break;
                default:
                    throw new IllegalArgumentException("Unsupported URI: " + uri);
            }
            //通知更改
            getContext().getContentResolver().notifyChange(uri, null);
            return count;
        }
}
```

7.2.4 ListView 界面的实现

分析完数据存储方式，下面我们按照程序运行的顺序来逐个对界面进行分析。首先是主界面，新建 MyContacts.java，继承于 ListActivity，因为需要实现的是一个 ListView 的界面。代码 2～4 行定义了菜单的 id 值，代码 8 行的作用是设置当前按键模式，以及当用户按键的时候会触发当前菜单中设置的快捷键功能。接着代码 15 行设置列表长按监听器，16～27 行查询联系人数据库中所有的数据，并将数据绑定 Adapter 中，最后为当前界面设置适配器 Adapter。MainActivity.java 文件的上半部分代码如下所示：

```
1   public class MyContacts extends ListActivity {
2       private static final int AddContact_ID = Menu.FIRST;
3       private static final int DELEContact_ID = Menu.FIRST + 2;
4       private static final int EXITContact_ID = Menu.FIRST + 3;
5
6       public void onCreate(Bundle savedInstanceState) {
7           super.onCreate(savedInstanceState);
8           setDefaultKeyMode(DEFAULT_KEYS_SHORTCUT);
9           // 为 intent 绑定数据
10          Intent intent = getIntent();
11          if (intent.getData() == null) {
12              intent.setData(ContactsProvider.CONTENT_URI);
13          }
14          // 设置菜单项长按监听器
15          getListView().setOnCreateContextMenuListener(this);
16          // 查询，获得所有联系人的数据
17          Cursor cursor = managedQuery(getIntent().getData(),
18                  ContactColumn.PROJECTION, null, null, null);
19
```

```
        // 注册每个列表表示形式 ：姓名 + 移动电话
        SimpleCursorAdapter adapter = new SimpleCursorAdapter(this,
                android.R.layout.simple_list_item_2, cursor, new String[] {
                        ContactColumn.NAME, ContactColumn.MOBILENUM },
                new int[] { android.R.id.text1, android.R.id.text2 });

        setListAdapter(adapter);
    }
```

7.2.5 创建菜单

作为一个 List 菜单，通常需要为其绑定监听器，分为长按和单击。当然，还包括任何界面都可以使用的创建菜单。

如下所示，代码 28～56 行为 menu 按键设置监听器，并为菜单的按键绑定监听器。当单击 menu 按键的时候，程序将创建两个菜单选项，一个是添加联系人，一个是退出，它们的快捷键分别是“a”和“d”。

代码 58～65 行设置列表选项的单击功能：查看联系人信息。67～75 行为长按选项监听器，当用户长按一个选项时，将弹出删除当前联系人菜单，用户可以通过这种方式删除指定的联系人。onCreateContextMenu 为长按菜单项触发的函数，而 onContextItemSelected 则为菜单项按键监听器。具体代码如下所示。

```
1   // 添加菜单选项
2       public boolean onCreateOptionsMenu(Menu menu) {
3           super.onCreateOptionsMenu(menu);
4           // 添加联系人
5           menu.add(0, AddContact_ID, 0, R.string.add_user).setShortcut('3', 'a').setIcon(R.drawable.add);
6
7           // 退出程序
8           menu.add(0, EXITContact_ID, 0, R.string.exit).setShortcut('4', 'd') .setIcon(R.drawable.exit);
9           return true;
10
11      }
12
13      // 处理菜单操作
14      public boolean onOptionsItemSelected(MenuItem item) {
15          switch (item.getItemId()) {
16          case AddContact_ID:
17              // 添加联系人
18              startActivity(new Intent(Intent.ACTION_INSERT, getIntent().getData()));
19              return true;
20          case EXITContact_ID:
21              // 退出程序
22              this.finish();
23              return true;
24          }
```

```
            return super.onOptionsItemSelected(item);
        }

        // 动态菜单处理
        // 单击的默认操作也可以在这里处理
        protected void onListItemClick(ListView l, View v, int position, long id)
    {
            Uri uri = ContentUris.withAppendedId(getIntent().getData(), id);
            // 查看联系人
            startActivity(new Intent(Intent.ACTION_VIEW, uri));
        }

        // 长按触发的菜单
        public void onCreateContextMenu(ContextMenu menu, View view,ContextMenuInfo menuInfo) {
            AdapterView.AdapterContextMenuInfo info;
            try {
                info = (AdapterView.AdapterContextMenuInfo) menuInfo;
            } catch (ClassCastException e) {
                return;
            }
            // 得到长按的数据项
            Cursor cursor = (Cursor) getListAdapter().getItem(info.position);
            if (cursor == null) {
                return;
            }

            menu.setHeaderTitle(cursor.getString(1));
            // 添加删除菜单
            menu.add(0, DELEContact_ID, 0, R.string.delete_user);
        }

        // 长按列表触发的函数
        @Override
        public boolean onContextItemSelected(MenuItem item) {
            AdapterView.AdapterContextMenuInfo info;
            try {
                // 获得选中的项的信息
                info = (AdapterView.AdapterContextMenuInfo) item.getMenuInfo();
            } catch (ClassCastException e) {
                return false;
            }

            switch (item.getItemId()) {
            // 删除操作
            case DELEContact_ID: {
                // 删除一条记录
                Uri noteUri = ContentUris.withAppendedId(getIntent().getData(),info.id);
                getContentResolver().delete(noteUri, null, null);
```

```
73                          return true;
74                  }
75              }
76              return false;
77          }
78  }
```

7.2.6 实现界面查看

跟主页面相关的两个界面分别是查看界面和编辑、添加界面。新建 ContactView 继承自 Activity，用于实现查看功能。在 onCreate 中根据传递过来的 UIR 地址获取联系人的信息，并将信息逐个显示到对应的界面位置中。然后，设置 menu 菜单，如代码 50～78 行所示。可以看到，这里设置了五个菜单项，第一个为删除当前联系人，调用 deleteContact()函数删除当前的联系人，接着返回主页面；第二个为返回列表，关闭当前的 activity，返回到主界面；第三个为编辑联系人，也就是下面我们要介绍的页面；第四个、第五个分别为呼叫联系人和发信息，将调用系统程序进行拨号和发信息。这里要注意 Intent 的特殊写法。具体代码如下所示。

```
1   public class ContactView extends Activity {
2       // 姓名
3       private TextView mTextViewName;
4       // 手机
5       private TextView mTextViewMobile;
6       // 座机
7       private TextView mTextViewHome;
8       // 住址
9       private TextView mTextViewAddress;
10      // 电子邮箱
11      private TextView mTextViewEmail;
12
13      private Cursor mCursor;
14      private Uri mUri;
15      // 设置菜单的序号
16      private static final int REVERT_ID = Menu.FIRST;
17      private static final int DELETE_ID = Menu.FIRST + 1;
18      private static final int EDITOR_ID = Menu.FIRST + 2;
19      private static final int CALL_ID = Menu.FIRST + 3;
20      private static final int SENDSMS_ID = Menu.FIRST + 4;
21
22      public void onCreate(Bundle savedInstanceState) {
23          super.onCreate(savedInstanceState);
24          mUri = getIntent().getData();
25          this.setContentView(R.layout.viewuser);
26          // 初始化界面元素
27          mTextViewName = (TextView) findViewById(R.id.TextView_Name);
```

```
        mTextViewMobile = (TextView) findViewById(R.id.TextView_Mobile);
        mTextViewHome = (TextView) findViewById(R.id.TextView_Home);
        mTextViewAddress = (TextView) findViewById(R.id.TextView_Address);
        mTextViewEmail = (TextView) findViewById(R.id.TextView_Email);

        // 获得并保存原始联系人信息
        mCursor = managedQuery(mUri, ContactColumn.PROJECTION, null, null, null);
        mCursor.moveToFirst();
        if (mCursor != null) {
            // 读取并显示联系人信息
            mCursor.moveToFirst();

            mTextViewName.setText(mCursor.getString(ContactColumn.NAME_COLUMN));
            mTextViewMobile.setText(mCursor.getString(ContactColumn.MOBILENUM_COLUMN));
            mTextViewHome.setText(mCursor.getString(ContactColumn.HOMENUM_COLUMN));
            mTextViewAddress.setText(mCursor.getString(ContactColumn.ADDRESS_COLUMN));
            mTextViewEmail.setText(mCursor.getString(ContactColumn.EMAIL_COLUMN));
        } else {
            setTitle("错误信息");
        }
    }

    // 添加菜单
    public boolean onCreateOptionsMenu(Menu menu) {
        super.onCreateOptionsMenu(menu);
        // 返回
        menu.add(0, REVERT_ID, 0, R.string.revert).setShortcut('0', 'r')
                .setIcon(R.drawable.listuser);
        // 删除联系人
        menu.add(0, DELETE_ID, 0, R.string.delete_user).setShortcut('0', 'd')
                .setIcon(R.drawable.remove);
        // 编辑联系人
        menu.add(0, EDITOR_ID, 0, R.string.editor_user).setShortcut('0', 'd')
                .setIcon(R.drawable.edituser);
        // 呼叫用户
        menu.add(0, CALL_ID, 0, R.string.call_user)
                .setShortcut('0', 'd')
                .setIcon(R.drawable.calluser)
                .setTitle(
                        this.getResources().getString(R.string.call_user)
                                + mTextViewName.getText());
        // 发送短信
        menu.add(0, SENDSMS_ID, 0, R.string.sendsms_user)
                .setShortcut('0', 'd')
                .setIcon(R.drawable.sendsms)
                .setTitle(    this.getResources()
                        .getString(R.string.sendsms_user)
                        + mTextViewName.getText());
```

```
            return true;
    }

    public boolean onOptionsItemSelected(MenuItem item) {
        switch (item.getItemId()) {
        // 删除
        case DELETE_ID:
            deleteContact();
            finish();
            break;
        // 返回列表
        case REVERT_ID:
            setResult(RESULT_CANCELED);
            finish();
            break;
        case EDITOR_ID:
            // 编辑联系人
            startActivity(new Intent(Intent.ACTION_EDIT, mUri));
            finish();
            break;
        case CALL_ID:
            // 呼叫联系人
            Intent call = new Intent(Intent.ACTION_CALL, Uri.parse("tel:"
                    + mTextViewMobile.getText()));
            startActivity(call);
            break;
        case SENDSMS_ID:
            // 发短信给联系人
            Intent sms = new Intent(Intent.ACTION_SENDTO, Uri.parse("smsto:"
                    + mTextViewMobile.getText()));
            startActivity(sms);
            break;
        }
        return super.onOptionsItemSelected(item);
    }

    // 删除联系人信息
    private void deleteContact() {
        if (mCursor != null) {
            mCursor.close();
            mCursor = null;
            getContentResolver().delete(mUri, null, null);
            setResult(RESULT_CANCELED);
        }
    }
}
```

7.2.7 添加一个标识变量

ContactEditor 界面是添加和编辑公用的界面，因此为了区分这两种情况，我们需要使用一个标志变量来区分。以下代码 11 行声明的 mState 变量即为标志变量，它可以取 STATE_EDIT 和 STATE_INSERT 两个值。

该界面的初始化函数 onCreate()最为关键，因为在这里需要决定如何显示。代码 26～47 行根据 Intent 传递的 action 值判断当前处于编辑还是新建状态，并将状态保存到 mState 中，以便于接下去的一系列操作。实际上，添加操作分为两步，第一步是当进入到这个界面的时候，已经新建了一个只包含 id 信息的数据，第二步是对这个数据进行更新操作。因此在代码 83～108 行都是从数据库取得相应联系人的信息，然后填充到指定的文本框。

代码 58～82 行分别设置了两个按钮的监听器，单击“保存”按钮时需要判断是否输入了东西，如果没有输入，则将原来的记录删除，否则更新数据。单击“取消”按钮则直接删除当前的数据，并关闭当前页面。具体代码如下：

```
1   public class ContactEditor extends Activity {
2           // 标志位常量，用于标记当前是新建状态还是编辑状态
3           private static final int STATE_EDIT = 0;
4           private static final int STATE_INSERT = 1;
5           //
6           private static final int REVERT_ID = Menu.FIRST;
7           private static final int DISCARD_ID = Menu.FIRST + 1;
8           private static final int DELETE_ID = Menu.FIRST + 2;
9
10          private Cursor mCursor;
11          private int mState; // 当前处于新建状态还是编辑状态的标志位变量
12          private Uri mUri;
13          // 界面元素
14          private EditText nameText;
15          private EditText mobileText;
16          private EditText homeText;
17          private EditText addressText;
18          private EditText emailText;
19          // 按键
20          private Button okButton;
21          private Button cancelButton;
22
23          public void onCreate(Bundle savedInstanceState) {
24                  super.onCreate(savedInstanceState);
25
26                  final Intent intent = getIntent();
27                  final String action = intent.getAction();
28                  // 根据 action 的不同进行不同的操作
29                  // 编辑联系人
30                  if (Intent.ACTION_EDIT.equals(action)) {
```

```
            mState = STATE_EDIT;
            mUri = intent.getData();
        } else if (Intent.ACTION_INSERT.equals(action)) {
            // 添加新联系人
            mState = STATE_INSERT;
            mUri = getContentResolver().insert(intent.getData(), null);
            if (mUri == null) {
                finish();
                return;
            }
            setResult(RESULT_OK, (new Intent()).setAction(mUri.toString()));
        }
        // 其他情况，退出
        else {
            finish();
            return;
        }
        setContentView(R.layout.editorcontacts);
        // 初始化界面文本框
        nameText = (EditText) findViewById(R.id.EditText01);
        mobileText = (EditText) findViewById(R.id.EditText02);
        homeText = (EditText) findViewById(R.id.EditText03);
        addressText = (EditText) findViewById(R.id.EditText04);
        emailText = (EditText) findViewById(R.id.EditText05);
        // 初始化按键
        okButton = (Button) findViewById(R.id.Button01);
        cancelButton = (Button) findViewById(R.id.Button02);
        // 设置确定按键监听器
        okButton.setOnClickListener(new OnClickListener() {
            public void onClick(View v) {
                String text = nameText.getText().toString();
                if (text.length() == 0) {
                    // 如果没有输入东西，则将原来的记录删除
                    setResult(RESULT_CANCELED);
                    deleteContact();
                    finish();
                } else {
                    // 更新数据
                    updateContact();
                }
            }
        });
        // 设置取消按钮监听器
        cancelButton.setOnClickListener(new OnClickListener() {
            public void onClick(View v) {
                // 不添加记录，也不保存记录
                setResult(RESULT_CANCELED);
                deleteContact();
```

```
                        finish();

                    }
            });
            // 获得并保存原始联系人信息
            mCursor = managedQuery(mUri, ContactColumn.PROJECTION, null, null, null);
            mCursor.moveToFirst();
            if (mCursor != null) {
                // 读取并显示联系人信息
                mCursor.moveToFirst();
                if (mState == STATE_EDIT) {
                    setTitle(getText(R.string.editor_user));
                } else if (mState == STATE_INSERT) {
                    setTitle(getText(R.string.add_user));
                }
                String name = mCursor.getString(ContactColumn.NAME_COLUMN);
                String moblie = mCursor.getString(ContactColumn.MOBILENUM_COLUMN);
                String home = mCursor.getString(ContactColumn.HOMENUM_COLUMN);
                String address = mCursor.getString(ContactColumn.ADDRESS_COLUMN);
                String email = mCursor.getString(ContactColumn.EMAIL_COLUMN);
                // 显示信息
                nameText.setText(name);
                mobileText.setText(moblie);
                homeText.setText(home);
                addressText.setText(address);
                emailText.setText(email);
            } else {
                setTitle("错误信息");
            }
        }
```

7.2.8 设置菜单

下面设置 menu 菜单，以下代码 1～17 行分两种情况为界面添加菜单，代码 19～39 行分别为菜单的按键设置功能。当为编辑模式时，创建的菜单包含“返回”和“删除联系人”两个选项。当为新建模式时，只有一个“返回”选项。

之后的代码实现删除和更新功能，主要是调用新建的 ContentProvider 的 delete()和 update()函数。

```
1       // 菜单选项
2       public boolean onCreateOptionsMenu(Menu menu) {
3           super.onCreateOptionsMenu(menu);
4           if (mState == STATE_EDIT) {
5               // 返回按钮
6               menu.add(0, REVERT_ID, 0, R.string.revert).setShortcut('0', 'r')
7                       .setIcon(R.drawable.listuser);
```

```
                // 删除联系人按钮
                menu.add(0, DELETE_ID, 0, R.string.delete_user)
                        .setShortcut('0', 'f').setIcon(R.drawable.remove);
            } else {
                // 返回按钮
                menu.add(0, DISCARD_ID, 0, R.string.revert).setShortcut('0', 'd')
                        .setIcon(R.drawable.listuser);
            }
            return true;
        }

        // 菜单处理
        @Override
        public boolean onOptionsItemSelected(MenuItem item) {
            switch (item.getItemId()) {
            // 删除联系人
            case DELETE_ID:
                deleteContact();
                finish();
                break;
            // 删除刚创建的空联系人
            case DISCARD_ID:
                cancelContact();
                finish();
                break;
            // 直接返回
            case REVERT_ID:
                finish();
                break;
            }
            return super.onOptionsItemSelected(item);
        }

        // 删除联系人信息
        private void deleteContact() {
            if (mCursor != null) {
                mCursor.close();
                mCursor = null;
                getContentResolver().delete(mUri, null, null);
                nameText.setText("");
            }
        }

        // 丢弃信息
        private void cancelContact() {
            if (mCursor != null) {
                deleteContact();
            }
```

```
                setResult(RESULT_CANCELED);
                finish();
        }

        // 更新变更的信息
        private void updateContact() {
                if (mCursor != null) {
                        mCursor.close();
                        mCursor = null;
                        ContentValues values = new ContentValues();
                        values.put(ContactColumn.NAME, nameText.getText().toString());
                        values.put(ContactColumn.MOBILENUM, mobileText.getText().toString());
                        values.put(ContactColumn.HOMENUM, homeText.getText().toString());
                        values.put(ContactColumn.ADDRESS, addressText.getText().toString());
                        values.put(ContactColumn.EMAIL, emailText.getText().toString());
                        // 更新数据
                        getContentResolver().update(mUri, values, null, null);
                }
                setResult(RESULT_CANCELED);
                finish();
        }
}
```

最后配置 AndroidManifest.xml 文件，代码如下所示。其中需要注意声明权限以及查看、编辑、新建界面的部分，里面使用了 data 标签。

```
<?xml version="1.0" encoding="utf-8"?>
<manifest xmlns:android="http://schemas.android.com/apk/res/android"
    package="com.example.contact"
    android:versionCode="1"
    android:versionName="1.0" >

    <uses-sdk
        android:minSdkVersion="8"
        android:targetSdkVersion="17" />

    <!-- 声明打电话、收发短信的权限 -->
    <uses-permission android:name="android.permission.CALL_PHONE" />
    <uses-permission android:name="android.permission.SEND_SMS" />
    <uses-permission android:name="android.permission.RECEIVE_SMS" />

    <application
        android:allowBackup="true"
        android:icon="@drawable/ic_launcher"
        android:label="@string/app_name"
        android:theme="@style/AppTheme" >

        <!-- 声明 contentProvider -->
        <provider
```

```
                android:name="com.example.contact.ContactsProvider"
                android:authorities="com.example.provider.ContactsProvider" />

            <!-- 主界面 -->
            <activity
                android:name=".MyContacts"
                android:label="@string/app_name" >
                <intent-filter>
                    <action android:name="android.intent.action.MAIN" />
                    <category android:name="android.intent.category.LAUNCHER" />
                </intent-filter>
            </activity>

            <!-- 编辑、新建界面 -->
            <activity
                android:name=".ContactEditor"
                android:label="@string/editor_user" >
                <intent-filter>
                    <action android:name="android.intent.action.EDIT" />
                    <category android:name="android.intent.category.DEFAULT" />
                    <data android:mimeType="vnd.android.cursor.item/vnd.guo.android.mycontacts" />
                </intent-filter>
                <intent-filter>
                    <action android:name="android.intent.action.INSERT" />
                    <category android:name="android.intent.category.DEFAULT" />
                    <data android:mimeType="vnd.android.cursor.dir/vnd.guo.android.mycontacts" />
                </intent-filter>
            </activity>

            <!-- 查看界面 -->
            <activity
                android:name="com.example.contact.ContactView"
                android:label="@string/view_user" >
                <intent-filter>
                    <action android:name="android.intent.action.VIEW" />
                    <category android:name="android.intent.category.DEFAULT" />
                    <data android:mimeType="vnd.android.cursor.item/vnd.guo.android.mycontacts" />
                </intent-filter>
                <intent-filter>
                    <category android:name="android.intent.category.DEFAULT" />
                    <data android:mimeType="vnd.android.cursor.dir/vnd.guo.android.mycontacts" />
                </intent-filter>
            </activity>

        </application>

</manifest>
```

7.3 知识拓展

上面的 AndroidManifest.xml 中使用到了 intent-filter 中的 data 标签，该标签包含的属性有 scheme、host、port、path、pathPrefix 和 pathPattern，这些属性的作用都是用来匹配 uri，书写形式为 scheme://host:port/pathorpathPrefixpathPattern。其中比较关键的一个属性是最后的 pathorpathPrefixorpathPattern，path 用来匹配完整的路径，如：http://example.com/blog/abc.html，这里将 path 设置为/blog/abc.html 才能够进行匹配；pathPrefix 用来匹配路径的开头部分。拿上面的 URI 来说。这里需要说一下匹配符合与转义。匹配符号："*"用来匹配 0 次或更多，如"a*"可以匹配"a""aa""aaa"……；"."用来匹配任意字符，如"."可以匹配"a""b""c"……。因此".*"就是用来匹配任意字符 0 次或更多，如".*html"可以匹配"abchtml""chtml""html""sdf.html"……；转义：因为当读取 xml 的时候，"\"是被当做转义字符的（当它被用作 pathPattern 转义之前），因此这里需要再次转义，读取 xml 是一次，在 pathPattern 中使用又是一次。如："*"这个字符就应该写成"*"，"\"这个字符就应该写成"\\\\"。

7.4 本章小结

本章介绍了如何开发一个简单的通讯录，从通讯录的添加、编辑、删除、查询这几个方面讲述了通讯录的开发过程。本章使用了 ContentProvider 进行数据操作，包括如何实现回调函数，如何调用，如何在 AndroidManifest.xml 文件中注册。最后介绍了 intent-filter 中 data 标签的使用。